Charles Hoare

The Slide Rule and How to Use It

Charles Hoare

The Slide Rule and How to Use It

ISBN/EAN: 9783337379704

Printed in Europe, USA, Canada, Australia, Japan

Cover: Foto ©berggeist007 / pixelio.de

More available books at **www.hansebooks.com**

THE SLIDE RULE

AND HOW TO USE IT

CONTAINING

FULL, EASY, AND SIMPLE INSTRUCTIONS TO PERFORM ALL BUSINESS CALCULATIONS WITH UNEXAMPLED RAPIDITY AND ACCURACY.

BY CHARLES HOARE, C.E.

AUTHOR OF "MENSURATION MADE EASY," ETC., ETC.

WITH A SLIDE RULE IN TUCK OF COVER

Seventh Edition

Capio Lumen

LONDON

CROSBY LOCKWOOD AND SON

7, STATIONERS' HALL COURT, LUDGATE HILL

1896

PREFACE.

To those who have once acquired a knowledge of the capabilities of the Slide Rule, it is ever a matter of surprise that an instrument combining such unexampled rapidity, ease, and accuracy in performing all ordinary business calculations, should be so little known. By its assistance the drudgery of computation is avoided, and the time and trouble expended on mere arithmetical workings proved to be a waste of effort ; in fact, its aid mentally may safely be compared with the advantages derived from mechanical appliances in ameliorating the wear and tear of manual labour.

The intellect remains unfettered by tedious processes, for the statement of each question, the operation and the result, are simultaneous and apparent in their connection. The laws that govern its operations are few and simple, and easily understood ; and the curiosity of the uninitiated may be stimulated by learning, that on an instrument as portable as a pocket-book we have the whole gamut of numbers ; and that whether as a means for self-instruction or advancement, for unsurpassed utility in business, or for profitable amusement, its study is well rewarded in its capabilities for varied application. Scientific men estimate its value, the man of business would soon appreciate its assistance, and it will be well for the practical mechanic when he learns how to employ it intelligently, instead of carrying it in his pocket, yet unable to avail himself of its extraordinary powers.

The disuse of the Slide Rule in ordinary calculations, in face of its proved capabilities, suggested the idea that either its construction, or the method of teaching, or perhaps both, might be capable of amendment. The adept may smile at the proposal to modify an instrument already simple enough to him, but there is evidence that, to make it available to many, it must first be made easy to all; for, generally speaking, its use has to be acquired by self-teaching, and if the professed instructions be not clear enough to pilot the beginner through the *seeming* difficulties of a new study, they are useless.

Sufficient introductory matter, and ample explanation, are needed to familiarise the student with the subject and its advantages. Treatises have been printed by the score, but Bevan, Woolgar, and other scientific writers, are scarcely before the public; while some are above, and many below general comprehension. To be popular, such matter must be plain. Abler pens might have invested the subject with greater interest; my aim has been simplicity of method and arrangement. Through the liberality of the present publishers of Weale's useful series, I have been unrestricted in space and detail; and believing that earnestness and accuracy may be accepted in lieu of higher pretensions, I trust that the large amount of information embodied in the work will prove acceptable and useful.

CHARLES HOARE.

CONTENTS.

Part V.—Commercial Arithmetic.

Part VI. Scientific Readings by Slide Rule.

THE SLIDE RULE.

INTRODUCTORY AND EXPLANATORY REMARKS.

THE combinations of the Slide Rule, like the elementary processes in Arithmetic, are few and simple, but their application is almost unlimited. Its action being mechanical, the working can happily be illustrated *without written rules*, in lieu of which, a copy of the position of the lines and figures, in fact, a diagram of the statement for each class of operations, is given, and the directions fully detailed in No. 6 of these Notes. The following memoranda, necessarily ample, are descriptive merely, enforcing no tax upon the memory after the explanations of the lines, numbers, and divisions on the Rule are clearly comprehended. No pains have been spared to render these preliminary instructions as concise and clear as possible. The successful practice of all that follows depends upon their being thoroughly *understood;* in such hands the Slide Rule is an intelligible and powerful instrument ;—in others a mere tool.

1st. All numbers and divisions are to be read decimally, for all the spaces are, or are supposed to be, divided and subdivided into tens and tenths ; the visible marks may describe fifths, or halves—these are still equal parts of ten. Where the spaces do not admit of subdivision, the proportions of $\frac{1}{4}$, $\frac{1}{2}$, $\frac{3}{4}$ must be estimated ; and when the eye grows accustomed to the scale, with a little practice, tenths of a division may be judged with great accuracy.

2nd. The figures on the Rule are engraved simply as 1, 2, 3, &c. ; out these numbers are arbitrary, and *any* required value may be assigned ; thus, a 2 may be called 2 or 20, or 200 ; if it is borne in mind that the whole line is affected during that operation.

Ex.: A 2 being called 20 the 3 is 30, and so on throughout the Scale on that line, but *different lines* may bear different values if the proportions are maintained, a simple case will illustrate this point.

R

Ex.: If 2 bushels cost 20s., what will 6½ bushels and also 65 bushels cost ?

The statement on the Rule would be—

A Set 2 bushels ·   6,5 bushels   and 65 bushels
B To 20 shillings   cost 65s.   cost 650 shillings

Here we have given the 2 *on the Slide* a tenfold value, viz. : 20s., which extends to the 65 and the 650.

3rd. The ORDINARY Slide Rule consists of four lines, viz. :—

A
B } Being on the Slide, the edge of B working with A.
C }   „   „   „   C   „   D.
D

This arrangement, though compact enough, is puzzling to the beginner, giving an unnecessary appearance of intricacy to a very simple instrument.

In the form of rule recommended and adopted by the Author, the pairs of lines

$\frac{A}{B}$ and $\frac{C}{D}$ } *are separated*, each pair having its own office, instead of a combined and complex operation. The following explanation of the uses and relation of these lines and slides is greatly assisted by an inspection of the Rule at the time.

$\frac{A}{B}$ Lines ; will be found exactly alike in numbers and divisions, and are, therefore, in direct simple proportion, and *all* such questions may be solved by them. When closed they stand thus :—

A 1   2   3   4 &c., &c.
B 1   2   3   4 &c., &c.

but if 1 in the Slide is projected to 2 on A, it will be seen that the ratio of $\frac{2}{1}$ runs throughout.

A 1   2   4   6 &c.
B   1   2   3 &c.

and so of any proportion we choose to make between $\frac{A}{B}$ lines.

$\frac{C}{D}$ Lines are relatively different, the effect being that all numbers and divisions on C are the square or self-multiple of the numbers on D, and consequently the numbers and divisions on D are the *square roots* of those on C, thus—

C 1   4   9   16 &c., &c.
D 1   2   3   4

Questions of Area or Solidity—proportions that are as to the square or square root of numbers—are solved by them.

4th. The reading or valuing of the figures and divisions on the several lines requires a little practice; the principle is simple enough. It has been shown that a numerical value being assigned to any figure, the rest of the line is determined by it, thus,

if the left	the centre	the right
is 1	is 10	is 1000
	or	
is 100	is 1000	is 10,000

the intermediate numbers and divisions follow in common notation; for between

				Subdivisions.
1 and	10	the prime divisions are Units		Tenths.
10 and	100	,,	Tens	Units.
100 and	1,000	,,	Hundreds	Tens.

It is here apparent that the lines *repeat*, at pleasure, and are therefore infinite, either in the ascending or descending scale. Had the numbering on the Rule been made absolute, its operations would have been limited, and valueless for questions above or below such stated quantities; the line D reads and repeats precisely in the same manner. It should be particularly noted when the squares and roots of numbers are in question, the true relation of C and D lines must be considered; thus, 2 on D is the square root of 4 on C, as 20 is of 400, *but not of* 40, which must be sought under the figure 4 on the right half of C, below which we find the correct square root of $40 = 6.32$.

C 1	4	40 right hand 4 on C
D 1	2	6.32 square root of 40

5th. The facility afforded by the Slide Rule in complex operations is greatly owing to the use of tabulated constants, the results of previous calculation, by the use of which fixed numbers all *similar* questions may be instantly solved; in Slide Rule practice these constants are *called* "Gauge Points," and are denoted by the abbreviation G. P., and any number in the body of this book so indicated is the constant of the question; but as substances vary in gravity, materials in strength, and bodies in dimensions; gauge points have to be found for each, and, when found, the labour saved by using them is almost incredible; sometimes their derivation is obvious, at others involved, but there is no mystery about them, and the beginner is not compelled to trace the nature of each before employing it; this can be done at leisure.

6th. To avoid the tedium of long printed directions for each working, which seriously divert the attention from the operation, the formula for stating each question will now be explained. It has been shown that the pairs of lines have different offices, and that the *proper pair* must always be employed. Great care has been taken to notify these, so that when $\frac{A}{B}$ or $\frac{C}{D}$ appear or are prefixed, that pair only is to be used. We will now take a diagram or formula from the body of the book; next show its full meaning; and then the simple working of the question.

See Measure of Capacity, Page 25.

C   Length in feet.   Answer in cubic yards.
D   G P 5.196,   Mean square feet.

FULL MEANING.

C Set the given length in feet   Find the answer in cube yards
    on C.   on C.
D To the gauge point 5.196   Above the M. square in feet
    on D.   on D.

SIMPLE WORKING.

Ex.: Required the cubic yards excavated, when the length is 100 feet, and the width and depth 8 feet each.

C   100   237 cube yards, answer
D   5.196   8

The formula is intended to show invariably the lines to be used, and the relative positions of the terms; *not* their *actual places* on the Rule, for the latter vary with their values.

It is necessary always to note how the terms of the question are placed, and to follow the reading of the statement, viz:—First, look what numbers (or constants if given) are to be set, and *after that*, for the result; as

A   1st. { Set ?   Then Below ?
B   { To ?   Find answer.

These *positions* may vary, as

A Then Below ?
B   Find answer   1st. { Set ?
                                  { To ?

but in every case the directions are clear, and by the formulæ under **any head all similar** questions may be unhesitatingly worked by merely substituting the given numbers or dimensions for those in the examples.

Before putting these directions into practice, it may be as well to recapitulate their chief points.

1st. That the Rule is numbered, and divided decimally.

2nd. That any value may be assigned to a number, if carried throughout that line.

3rd. That the proper lines $\frac{A}{B}$ or $\frac{C}{D}$ are always to be used, and the question correctly stated, as in the formula

4th. That the reading or valuing of the numbers be well understood.

5th. That the correct G.P. is selected, and used as prescribed.

6th. That the positions in the formula do not necessarily imply the actual *places* of the numbers on the Rule, but are intended to show the exact order of stating *any similar* question.

ABBREVIATIONS.

G. P.	Denotes "gauge point." If the number is given, it is to be employed; but if not given, the proper one is to be selected.
a	(a) affixed to any number in a formula, shows the answer, or place of the answer.
A 1 or B 1, &c.	Shows that 1 (unity) on that slide or line is to be set to some given number.
×	Multiplier, or the *place* of a multiplier, or the office of the gauge point used.
÷	Divisor, with the same uses.
A B or C D	Prefixed to any formula, shows which lines are to be employed.
B. Invt^d C. Invt^d	The Slide has sometimes to be *inverted*; and in any operation where this is required, the letters Invt^d indicate it, and must be attended to.

THE ARRANGEMENT OF THE SLIDE RULE,

Issued with " Weale s Scientific Series."

T‚HIS, the simplest form of the Slide Rule, is strongly recommended by the author for all ordinary purposes of calculation; the separation of the usual lines A B, C D, needlessly connected on the common Rule, removes the appearance of intricacy so puzzling to the beginner; and every formula in this work is adapted to a *single line and slide*, viz. :—

Either $\dfrac{A}{B}$ or $\dfrac{C}{D}$

and in each case the letters prefixed to the Example show unhesi-tatingly which pair is to be employed.

The Rule in this form being used for calculations only, and not like the common rule—half tool and half instrument—the utmost care has been expended on its construction, and its accuracy guaran-teed.

LESSONS FOR PRACTICE.

THE correct reading of the numbers and divisions being of primary importance, so as to be able *at sight* to find or assign any required value, practice in this is essentially necessary, and the following method will greatly assist the beginner; A, B, and C lines being *exactly alike*, it is sufficient to take one of them, say A.

	Left Half.	*Right Half.*
A 1	2 3 4 5 6 7 8 9 10	20 30 40 50 60 70 80 90 100

In the first reading, omit all divisions.

Then repeat, introducing the chief divisions.

A 1. { Read units and *tenths* up to............ 10 Read tens and *units* up to.......... 100

Repeat, noting any subdivisions, as halves or fifths.

A 1. { Read units, tenths, and any subdivisions ... 10 Read tens, units, and any subdivisions........100

Now repeat, trying to estimate $\frac{1}{4}$, $\frac{1}{2}$, $\frac{3}{4}$, where undivided.

A 1. .... 10 100

Commence the line at 100, and estimate subdivisions.

A 100. { The *divisions* are tens............1,000 The *divisions* are hundreds 10,000

Having practised these scales sufficiently, try and make out the following numbers correctly :—

	Left Half.	*Right Half.*
A 1.	2·2 2·7 3·6 4·9 7·5	12 15 27 34 50 63 75
	144 225 337 450 725	3550 4300 5750 7500 8750

D Although D, a single line, is to be read, valued, and *repeated* precisely as A, B, and C, yet affording more space, the divisions and subdivisions are carried further.

To confirm the above practice, let the learner test his accuracy by some *printed* tables; and as $\frac{C}{D}$ form of themselves lines of squares and roots (No. 5), proceed leisurely, and compare the Rule with the book; having the actual figures to guide and check the reading, let him try carefully to *estimate* the reading of undivided spaces, seeing how closely he can approximate to the 1st, then the 2nd place of decimals.

If books are handy, open a Table of Diameters and Areas; and to form the same on the Rule, move the slide C gently till number 7 on C coincides with number 3 on D, by which a Table of Areas and Diameters is instantly made.

C . . . . The areas of circles on C . . . . Set 7

D . . . . To diameters   on D . . . . To 3

then proceed with the comparison, till the book is used as a check, and not wholly as a guide. There is no preliminary practice so good as this, as it gives the learner confidence in the Rule and in himself. The perception of the power placed within his reach soon gives interest to each task, for the matter contained in *folios of such printed tables* lies in the instrument, available to all who have pride and patience enough to master it.

PART I.

INSTRUMENTAL ARITHMETIC.

THE Notes on Slide Rule Arithmetic will be found page 12; if mixed with the Formula and Examples, they might divert attention from the simple working.

---

1. MULTIPLICATION on A and B. (Note 1).

Ex.: Multiply 9 by 3 = 27.

A To 3   Find 27 answer
B Set 1 on the Slide   Above 9 on the Slide

2. DIVISION on A and B.

Ex.: Divide 27 by 9 = 3.

A Find 3 answer   To 27
B Above 1 on the Slide   Set 9

3. PROPORTION (Direct) on A and B. (Note 2).

Ex.: As 2 : 6 :: 12 : 36.

A To 6   Find 36 answer
B Set 2 on the Slide   Above 12

When several proportions are required.

Ex.: As 18 : 252 :: 3', 4, 5, and 6 shares.

A Find 42   56   70   84 answer   To 252
B Above 3   4   5   6   Set 18

4. PROPORTION (Inverse) on A with the Slide *inverted*. (Note 3).

Ex.: As 6 : 4 :: 12 : 2.

A   Find   2   answer   To 4
S. Invt^d Above 12   Set 6

5. Squares and Roots of Numbers on C and D.

C Set 1 (C the Slide) is then a line of square numbers.
D To 1 D line shows the square roots of all Nos. on C.

Ex. : Find the square roots of 4, 9, 25, and 36.

| C Set 1 | 4 | 9 | 25 | 36 |
| D To 1 | 2 | 3 | 5 | 6 answers. |

6. Cubes and Roots on C and D.   (Note 4).

To cube a number or dimensions—*Ex.* : Cube 4.

C Set 4 Find 64 the cube
D To 1 Above 4

The given number on C may be set to 1 *or* 10, as the Slide sometimes over-runs the numbers on D.

Ex. : Required the cube root of 64.   (Invert the Slide to D).
Note 4.

(S I) Find 4 the root Set 64
D 4 To 10

7. 4th Power and Roots on C and D.   Note 5.

Ex. : Find the 4th power of 3.

C
D 1st. { Find 9 its 2nd power.   Find 81 the 4th **power**
{ Above 3   Then above 9

Ex. : Find the 4th root of 81.

C Then below 9   1st { Below 81.
D   Find 3 the 4th root of 81   { Find 9 its sq. root.

8. Mean Proportional, or Mean Square of Unequal Sides
on C and D.   See Note 6.

Ex. : Mean Square of 4 and 9.

C Set 4 the less No.   Below 9 the greater No.
D To 4 its like No.   Find 6 the mean.

9. Two Mean Proportionals, a double setting required.

Ex. Find two mean proportionals to 2 and 16.

1st, with Slide inverted.

(S. Invt^d) Set 16	4 }	are coincident numbers and the
D To 2	4 }	1st proportional.

2nd with Slide rectified.

C Set 4, 1st proportional   Below 16.
D To 4, its like No.   Find   8 the 2nd proportional.

Answer, 2   4   8   16.

10. To Multiply Squared Numbers.

In effect the same as Cubing. See No. 6 and Note 7.

Ex.: Multiply 3.5^2 by 4.

C Set 4   Find   49 the answer.
D To 1   Above 3.5

11. To Divide Squared Numbers C and D.

Ex. Divide 6^2 by 4.

C Set 4   Find 9 the answer.
D To 4   Above 6.

12. To Divide by a Squared Number, C and D.

Ex.: Divide 144 by 3^2.

C Find 16 answer   Set 144
D Abore 1   To   3

DECIMAL AND COMMON FRACTIONS.

13. Decimal Equivalents, or A and B.

Ex.: Convert the fraction ⅝ to a decimal.

A Find the .625 the dec. equivalent   Set 5 } or with any given
B Above   1 on the Slide   To 8 } fraction on the rule

14. DECIMALS TO COMMON FRACTIONS.

Ex. : Convert the decimal .625 to a common fraction.

A Set .625   Find $\left\{ \dfrac{5}{8} \right.$ (Note 8).

B To   1

15. RECIPROCALS on A and B.

Ex. : Find a divisor equal to the multiplier **4.**

A Below 1   To 4 ×

B Find .25 ÷ answer   Set 1.

Ex. : Find a multiplier equal to the divisor **5.**

A Below 1.   To 5 ÷

B Find .2 × answer   Set 1

16. COMMON FRACTIONS, valued in any denominations on **A** and **B**

Ex. : Find the value of $\frac{3}{8}$ of 20 shillings.

A Find   7.5 shillings   To 20 shillings

B Above 3   the numerator   Set 8 the denominator

Any fraction may be as easily reduced, as $\frac{9}{18} = \frac{3}{5}$ of 37.

A 22.5 answer   37.

B 3   5.

17. TO FIND THE VALUE OF DECIMAL PARTS OF ANY INTEGER.

Read the *whole line* A as a scale of decimal parts, thus :—

Beginning		Centre.	Tenths.		Right.
A .01	.02 .03 &c. to .1	.2	.3	.4 &c. to	1. whole

Any denomination being placed under 1, on the *right* of the Scale A, the decimal values of any *parts* on B are shown on A, or on B, find the value in parts of any decimal on A. (Note 9.)

NOTES ON INSTRUMENTAL ARITHMETIC.

NOTE 1.—*Multiplication.*—When 1 on the Slide is set to any number on A, *all* numbers assume that ratio, and the product of ANY number on B may be found on A without shifting the Slide.

Note 2.—*Proportion Direct.*—If more requires more, or less requires less, the question is one of Direct Proportion. The statement on the Rule may be varied thus :—

Let　　　　a　　b　　c　　d
Represent　2　　4　　6　　12　．　：　：　：　：

Then, by Slide Rule—

$$x = \frac{b \cdot c}{d}$$　　
$\begin{array}{ll} \text{A} & 2 \ x \\ \text{B} & 6 \ c \end{array}$　　
$\begin{array}{ll} 4 \ b \\ 12 \ d \end{array}$　or　
$\begin{array}{ll} c & 6 \\ x & 2 \end{array}$　　
$\begin{array}{ll} 12 \ d \\ 4 \ b \end{array}$

$$r = \frac{a \cdot d}{c}$$　　
$\begin{array}{ll} \text{A} & 4 \ x \\ \text{B} & 12 \ d \end{array}$　　
$\begin{array}{ll} 2 \ a \\ 6 \ c \end{array}$　or　
$\begin{array}{ll} d & 12 \\ x & 4 \end{array}$　　
$\begin{array}{ll} 6 \ c \\ 2 \ a \end{array}$

$$x = \frac{a \cdot d}{b}$$　　
$\begin{array}{ll} \text{A} & 6 \ x \\ \text{B} & 2 \ a \end{array}$　　
$\begin{array}{ll} 12 \ d \\ 4 \ b \end{array}$　or　
$\begin{array}{ll} a & 2 \\ x & 6 \end{array}$　　
$\begin{array}{ll} 4 \ b \\ 12 \ d \end{array}$

$$x = \frac{b \cdot c}{a}$$　　
$\begin{array}{ll} \text{A} & 12 \ x \\ \text{B} & 4 \ b \end{array}$　　
$\begin{array}{ll} 6 \ c \\ 2 \ a \end{array}$　or　
$\begin{array}{ll} b & 4 \\ x & 12 \end{array}$　　
$\begin{array}{ll} 2 \ a \\ 6 \ c \end{array}$

In Direct Proportion, the multipliers never appear on the same line, and are always in opposition,

As $\dfrac{A}{B}$　　$\begin{array}{c} \text{Answer} \\ \times \end{array}$　$\begin{array}{c} \times \\ \div \end{array}$　or　$\begin{array}{c} \times \\ \div \end{array}$　$\begin{array}{c} \text{Answer} \\ \times \end{array}$

Note 3.—*Proportion Inverse.*—If more requires less, or less requires more, the question is in Inverse Proportion.

There are two methods of stating such questions. I believe I recommend the simplest, in *reversing* the Slide, for the lines A B are thus changed from Direct to *Inverse* Proportion, when the statement at once becomes clear.

Ex.: As 6 : 4 : : 12 : 2.

A　　　　Set 6　　　　　　　Below 12.
(S. Inv^d)　To 4　　　　　　　Find　2 Answer.

It is only necessary for observation, but useful to note, that in all Inverse questions, if correctly stated, the products of the vertical numbers on the rule are equal.

$$\begin{array}{c} 6 \\ 4 \\ \hline 24 \end{array} \quad = \quad \begin{array}{c} 12 \\ 2 \\ \hline 24 \end{array}$$

Note 4.—*Cube Root.*—This tedious arithmetical process would be performed on the Slide Rule by mere inspection, the same as squares and roots, were C a *triple* line to D single; in fact, any roots and powers might be found by increasing the sections of C to the required power (the present double radius gives the 2nd, and by No. 7, the 4th, and 8th). In the absence of an arrangement, which would affect the simple form—squares and roots being sufficient for all general purposes—a little ingenuity must be substituted to find cube roots; we therefore *reverse the Slide*, and, setting the given number to 1 or 10 on D, have to look for the numbers or divisions exactly coinciding, which is the required cube root; as in the example—

$$64 \sqrt[3]{}$$

| S. Inv^d | Find the coincident } 4 { the cube root | 1st { Set 64. |
| D | Numbers ...... } 4 { of 64. | To 10. |

In this case it is easy and evident that no other similar numbers meet; but, sometimes, owing to the differing scales of C and D, and to one being inverted, a little patience is needed, to mark the intersecting numbers or divisions on } C Ivt^d the only case in Slide Rule } D, practice demanding it, but still wonderfully simple compared with arithmetic.

Note 5.—*4th Power and Root.*—To those acquainted with Evolution, the example given will suffice, for C and D forming lines of squares and roots, we can find the square root of the square root; and, consequently, the 4th power or root of any number by inspection.

Note 6.—*Mean Proportional, or Mean Square.*—This rule is important in its application to cubed work, for, *unless* the area of the section is found, the true mean square must be used in multiplying by the length, or serious error occurs. For instance—

$$4 \times 9 = 36 \sqrt{} \qquad = 6, \text{ the true mean.}$$
$$\text{whereas } 4 + 9 = 13 \div 2 \qquad = 6.5, \text{ the arithmetical mean.}$$

Wherever the mean square is demanded in this work, it must be found by the easy method given page 10.

| The effect of setting the 4 on C | Is to project the 9 on C |
| To its like No.   4 on D | To their sq. root 6 on D |

Note 7.—*Multiplying Squared Numbers.*—This is the same as in cubing; for the line D being already, by its proportions, equal to the *square* of its numbers on C,

$$\text{as } \quad \begin{array}{cccc} C & 1 & 4 & 9 \\ D & 1 & 2 & 3 \end{array} \quad \&c., \&c.$$

and the setting of any number on C ?
to D 1,

being to multiply *all* numbers on D by that number. If it represents the length of a solid, we have

Squared-up section × length, or the cube contents on C.

Note 8.—*Converting Decimal and Common Fractions.*—When unity (1 on B) is set to any decimal on A,

then A represents the numerators ⎰ of all corresponding
and B   ,,   ,,   denominators ⎱ common fractions,

from which the lowest term can be selected by inspection. For example—

A To .125 gives $\frac{1}{8}$ .25 gives $\frac{1}{4}$ .375 gives $\frac{3}{8}$, &c.
B Set 1.   1   1

But should no unit divisions coincide, the fraction may be found among the tens.

$$Ex.: \quad \frac{A}{B} \quad \frac{To}{Set} \quad \frac{.766}{1} \quad = \quad \frac{23}{30}$$

Note 9.—*Decimal Values.*—The construction of the Slide Rule, necessitating the reduction of all statements and readings to decimal expressions, facility in this operation is essential; but the examples given are so clear, and the methods so speedy and simple, even with the complex divisions and subdivisions of our standards, that a little practice removes every difficulty. It is recommended to compare the readings of the instrument with some *printed* decimal tables. A correct knowledge of the scale of divisions is, at the same time, acquired; and such lessons will be found to have many advantages over unchecked readings.

Ex.: Place *in succession*, under 1, on the right of A, the denominators of a £, a yard, a rod, an acre, or of *any* integer, and form decimal scale and value.

under
A Read Line of decimal parts on A+1 on the right.
Set

B	Only one denomination at a time.	Above any Divisions of the Integer	+20 shillings = 1 £.
B			36 inches	1 yard.
B			272 feet	1 rod.
B			160 rods	1 acre.
B			12 pence	1 shilling.
B			12 inches	1 foot.

The different processes in Slide Rule Arithmetic will be found applied in the Parts which follow; and reference made to these notes, where necessary, to explain any operation more fully than the general formula would permit.

PART II.

MENSURATION.

QUESTIONS in Mensuration are solved with remarkable facility by the Slide Rule; for not only are the tedious processes in finding the contents of solids avoided altogether, but, in the majority of cases, given (see Note 5) tabular results of great value are obtained. For greater convenience of reference, this subject is divided into—
1st, Linear; 2nd, Superficial; and 3rd, Solid Measurement.

1st. LINEAR MENSURATION.

1. DIAGONALS OF SQUARES ON A B:

A	Find the diagonals of all squares	1st	Set 99
B	Above the sides do.		To 70

Or on C and D :

C	Set 1		Below 2
D	To side of any square		Find diagonal

Ex.: The side of a square being 3.5, required its diagonal.

C	Set 1		Below 2
D	To 3.5		Find 4.95 diagonal

2. DIAGONALS OF CUBES ON C and D :

C	Set 1	Below 2		Below 3
D	To side of *any* cube	Find diag. of face		Find diag. of cube

Ex.: The side of a cube being 3.5, required the diagonals of its face and solid.

C	Set 1	Below 2		Below 3
D	To 3.5	Find 4.95 diag. of face		Find 6.06 diag. of cube

3. ONE MEAN PROPORTIONAL (on C and D).

C	Set the least number		Below the greater number
D	To the least number		Find the mean proportional

Ex.: Find a mean proportional to 4, and 9.

C   Set 4   Below 9

D   To 4   Find 6 the mean

4. Two Mean Proportionals (on C D).

(Inv^{td} Slide). Set the greater **No.**   * ⎧ Observe where any numbers

D   To the least   * ⎨ or divisions coincide, *they*

                                   ⎩ *will be* 1*st proportional.*

Then rectify the Slide.

C   Set * 1st proportional   Below the greater Number

D   To * 1st proportional   Find the 2nd proportional required

Ex.: Find two mean proportionals to 2 and 16.

Slide Inv^{td} Set 16 greater No.   4 will be found coincident and

D   To 2 least No.   4 is the 1st proportional

Then—

C   Set 4   Below 16

D   To 4   Find   8 the 2nd proportional

Answer—2 . 4 . 8 . 16.

5. Sides of Squares of equal Area (on A and B).

A
B   Tabular ⎧ Find sides of squares of equal area   Set 39
          ⎩ Above diameters of circles   To 44

A
B   Tabular ⎧ Find sides of squares of equal area   Set 11
          ⎩ Above circumferences of circles   To 39

6. Sides of Squares to be Inscribed (on A and B).

A
B   Tabular ⎧ Find sides of squares to be inscribed   Set 53
          ⎩ Above diameters of circles   To 75

A
B   Tabular ⎧ Find sides of squares to be inscribed   Set 45
          ⎩ Above circumferences of circles   To 200

7. Diameters of Circular Segments (on A and B).

A   Set the versed sine   Below the $\frac{1}{2}$ chord

B   To the $\frac{1}{2}$ chord   Find the supplement, which, added to
                              the v. sine = diameter

Ex.: The chord of a segment being 48, and the versed sine 18, required the diameter of the circle.

A   Set v. sine $=$ 18   Below 24 $=$ ½ chord
B   To ½ chord $=$ 24   Find   32 the Supplement

$$\left. \begin{array}{c} \text{Supplement (add) to v. sine} \\ 32 \quad + \quad 18 \end{array} \right\} = 50 \text{ the diameter required}$$

8. LENGTH OF ARC (n) DEGREES (on A and B).

Note.—Find the diameter by the last rule if necessary.

A   To 115   Below any given degrees
B   Set the diameter   Find the length of arc

9. CIRCUMFERENCE AND DIAMETER OF CIRCLE (on A and B).

A   Set 22   Line of circumferences
B   To   7   Line of diameters.

10. ELLIPSE CIRCUMFERENCE (on A and B).

A   Set 70   Find the circumference of any ellipse
B   To 45   Above the *sum* of the two diameters

11. ISOMETRICAL ELLIPSE (on C and D).

C   Set 1   Below 2.   Below 3.
D   To the Minor Axis   Find Isomet. Diam^r.   Find the Major Axis.

12. RIGHT-ANGLED TRIANGLES (on C and D).

To find the hypothenuse—

C   Set 1 Find its square.   Find its square (add) $\left\{ \begin{array}{l} \text{and below the} \\ \text{sum of squares} \end{array} \right.$

D   To 1 Above the 1st leg   Above the 2nd leg $\left\{ \begin{array}{l} \text{Find the hypothe-} \\ \text{nuse on D.} \end{array} \right.$

Ex.: The sides of R. A. triangle being 4 and 6, required the hypothenuse.

C   Set 1 Find   16 $+$ 36 $=$ 52 below which find
D   To 1 Above 1st leg 4   2nd leg 6   7.21 the hypothenuse.

To Find either leg.

The hypothenuse added to one leg $=$ Sum *

Then—

C   Set * sum   Below hypothenuse *less the same leg*
D   To * sum   Find the required leg.

13. OTHER TRIANGLES (on A and B).

Note.—By the use of the natural sines, &c., the sides and angles of triangles may be found on A and B; space cannot be given for more than the sines and differences of *full degrees*, to show their use, or for approximate working. Lines of sines, tangents, &c., can be laid down on a Slide, and by their aid and the line A, the necessity for any tables is avoided, as all trigonometrical questions that are solved by using logs and log sines are answered on the Rule.

TABLE of Natural Sines to full Degrees, with differences from which 10′ 20′ &c., may be found.

Deg. N.S.	Dif.	Deg. N.S.	Dif.	Deg. N.S.	Dif.	Deg. N.S	Dif.
1 .017		23 .391		45 .707		69 .933	
2 .035		24 .407		46 .719		70 .939	.006
3 .052		25 .423	.016	47 .731	.012	71 .945	
4 .070		26 .438		48 .743		72 —.951	
5 .087		27 .454		49 —.755		73 .956	.005
6 .104		28 .469		50 .766	.011	74 .961	
7 .122		29 —.485		51 .777		75 —.966	
8 .139		30 .5		52 —.788		76 .970	.004
9 .156	difference .017	31 .515	.015	53 .798		77 .974	
10 .174		32 .530		54 .809		78 —.978	
11 .191		33 —.545		55 .819	.010	79 .981	.003
12 .208		34 .559		56 .829		80 .984	
13 .225		35 .573		57 .838		81 .987	
14 .242		36 .588	.014	58 —.848		82 —.990	
15 .259		37 .602		59 .857	.009	83 .992	.002
16 .276		38 .616		60 .866		84 .994	
17 .292		39 .629		61 —.875		85 —.996	
18 .309		40 —.643		62 .883		86 .997	.001
19 .325		41 .656		63 .891	.008	87 .998	
20 .342		42 .669	.013	64 —.899		88 —.999	
21 .358		43 .682		65 .906		89 .999	
22 —.375		44 —.695		66 .913	.007	90 1. —	
				67 .920			
				68 —.927			

Ex.: The side of a triangle (whose opposite angle is 40° sine 643 from Table) equals 800 yards; required the side, the angle being 30° sine .5 in table.

A Find 622 yards required side   To 800 the yards in given side.
B Above ·5 the sine of 30″   Set ·643 the sine of 40°

14. REGULAR POLYGONS Linear proportions (on A and B).

			Sides.	3.	4.	5.	6.	7.	8.	9.	10.	11.	12.
Find Radius...............	On A	Set	56	70	74	1	C0	98	22	89	80	85
Above Len. of Side......	On B	To	97	99	87	1	52	75	15	55	45	44
Find Perpendicular	On A	Set	9	1	40	26	25	40	40	40	29	28
Above Len. of Side......	On B	To	31	2	58	30	23	33	29	26	17	15
Find Perpendicular ...	On A	Set	1	20	21	26	27	25	33	20	25	28
Above Radius	On B	To	2	28	26	30	30	27	35	21	26	29

Note.—When the constants for any polygon are set on A and B, the proportions named in the margin appear in tabular form on their respective lines for all figures of that kind.

Ex.: It is required to form a table of side and radius for octagons.

A is then a line of radii
B   ,,   ,,   sides

1st { Set 98 on A { Constants for
      { To 75 on B { octagons.

15. DIVIDING LINES INTO EQUAL PARTS on (A and B).

A Find length of part required   To the whole length
B Above 1   Set the No. of parts

2ND. MENSURATION OF AREAS.

1. SQUARE SUPERFICES (on A and B).

When dimensions in Feet   ×   Feet, the answer in *square feet.*
A To 1   Below the breadth
B Set length feet   Find square feet super

When dimensions in Feet   ×   Inches, the answer in sq. feet.
A To 12   Below the breadth in inches
B Set length feet   Find sq. feet super

When dimensions in Inches   ×   Inches, the answer in square feet.
A To 144   Below breadth in inches
B Set length inches   Find sq. feet super

When dimensions in Yards   ×   Yards, the answer in *square yards.*
A To 1   Below the breadth in yards
B Set length in yards   Find sq. yards super

When dimensions in Yards × Feet, the answer in square yards.
A To 3   Below the breadth in feet
B Set length in yards   Find sq. yards super

When dimensions in Feet × Feet, the answer in square yards.
A To 9   Below the breadth in feet
B Set length in feet   Find sq. yards super

When dimensions in Chains × Chains, the answer in *acres*.
A To 10   Below the breadth in chains
B Set length in chains   Find acres area

When dimensions in Perches × Perches, the answer in acres.
A To 160   Below the breadth in perches
B Set length in perches   Find acres area

When dimensions in Yards × Yards, answer in acres.
A To 4840   Below the breadth in yards
B Set length in yards   Find acres area

When dimensions in Yards × Yards, answer in perches.
A To 30.25   Below breadth in yards
B Set length in yards   Find perches area

Ex.: A sheet of plate glass is 7 feet long, and 34½ inches wide—required the sq. feet.

A To 12   Below 34 inches wide
B Set 7 feet long   Find 20⅛ sq. feet super

2. TRIANGLES (on A and B).

A Set length of bas   Find the area
B To 2   Above the perpendicular

3. REGULAR POLYGONS (on C and D).

	Sides.		3.	4.	5.	6.	7.	8.	9.	10.	11.	12.	16.	20.	24.
Find Areas ...	On C	Set	21	9	*43	65	58	310	500	490	150	100	500	500	410
Above Sides...	On D	To	7	3	5	5	4	8	9	8	4	3	5	4	3

Ex.: Required the area of a Pentagon, the side being 4 feet (and form a table of areas of Pentagons).

C Find 27.5 area   Set 43 } Constants for Penta-
D Above 4 the given sides   To 5* } gons.

4. TRAPEZOIDS (on A and B).

A Set the sum of Parallel sides   Find super. contents
B To 2   Above the breadth

5. *Circular Areas.*

6. DIAMETERS AND AREAS (on C and D).

C Set 7   Line of areas on C
D To 3   Line of diameters on D

Ex. Find the Areas to Diameters 5, 6, and 7.

C Set 7   Find 19.6, 28.3, 38.5 areas required
D To 3   Above 5,   6,   7 diameters given

7. DIAMETERS IN INCHES, *Areas in Square Feet* (on C and D).
C Set 11   Find areas in square feet on C
D To 45   Above any diameter in *inches* on D

8. CIRCUMFERENCES AND AREAS (on C and D).

C Set 23   Line of areas on C
D To 17   Line of circumferences on D

(Or on A and B).

A Set the radius   Find the area on A
B To 2   Above the circumference on B

9. AREAS OF CIRCLES AND INSCRIBED SQUARES (on A and B).

A Set 11   Find areas of circles on A
B To 7   Above areas of inscribed squares

Nos. 6 to 9, yield tabular results.

10. AREAS OF SECTORS OF CIRCLES (in) DEGREES (on C and D).
C Set 465   Below any given No. of Degrees
D To diameter of circle   Find square root of area

Close the Slide, and above the square root on D, find the square = area of sector on C.

(Or on A and B).

A Set the radius of circle   Find area of sector required
B To 2   Above the length of arc

Example of No. 10. The diameter of a circle is 20, the degrees of the sector 15; required the area.

C Set 465   Below 15°
D To 20   Find 3.62 the square root of area
                         = 13,09 area

11. SQUARE AND CIRCULAR AREAS COMPARED (on A and B).

A Set 70   Line of circular feet, inches or measures
B To 55   Line of square   ,,   ,,   ,,

Note.—The circular foot contains 113.0977 square inches.
      The square foot contains 183.346 circular inches.

12. ELLIPTICAL AREAS (on A and B).

A Set 55   Find area of ellipse on A
B To 70   Above the *product* of its diameters on B

Ex.: required the area of an ellipse, the diameters being 15 and 20 = 300.

A   55   235.6 area
B   70   300

or

A Set long. diameter   Find area
B To 1.273   Above short diameter

13. PARABOLIC AREAS (on A and B).

A Set the base   Find the area
B To 1.5   Above the altitude

14. CONICAL SURFACES (on A and B).

A Set the circumference of base   Find the curve surface
B To   2   Above the siant height

Or—

A   Set the diameter   Find the curve surface
B   To   .64   Above the slant height

Ex. of No. 14.   The diameter of the base of a cone is 5 inches, its slant height 18 inches; required the curve surface.

A   5. Diameter   141.3 Curve surface required
B   .64   18 Slant height

15. CONVEX OF SPHERES (on C and D).

C   Set 50   Find the convex surface
D   To   4   Above the diameter of any sphere

Ex.: The diameter of a sphere is 8; required its convex surface.

C   50   201 convex surface
D   4.   8 diameter

16. CONVEX OF SPHERICAL SEGMENTS (on A and B).

A   Set height of segment   Find convex of segment
B   To   1   Above circumference of sphere

Ex.: The circumference of a sphere being 25.1, and the height of a segment of it being 2, required the convex surface.

A   2 H   50.2 convex surface, answer
B   1   25.1 circumference

17. SURFACE OF REGULAR SOLIDS (on C and D).

Number of facets	4	6	8	12	20
			*		
Find the whole surface on C Set	85	24	170	330	78
Above the edge of facet on D To	7	2	7	4	3

Ex.: Required the surface of a regular solid of 8 sides, the edges of each facet being 4 inches.

C   *   170   55.5 inches surface
D   7   4 inch edge

18. Curve Surface of Cylinders (on A and B).

A   Set the diameter   Find the curve surface
B   To   .32   Above the height

Or—

A   Set the circumference   Find the curve surface
B   To   1   Above the height

Ex.: The diameter of a cylinder is 8 inches, and its height is 12 inches; required its curve surface.

A   8 inches   301.6 inches curve surface
B   .32   12   height

Note.—The greatest cube content of a cylinder bounded by the least surface is when the diameter is made equal to the required cube contents multiplied by 2.5465, and the cube root of the product taken; its depth being half such diameter.

3rd. MENSURATION OF SOLIDS.

---

1. Cubical Forms (on C. and D).

When the cube contents are required in the *same measure* as that in which the dimensions were taken.

C   Set the length   Find the cube contents
D   To 1   Above the mean square of side

Ex.: The length of a shaft being 30 feet, and its sides 5 × 4 feet; required the cube contents.

C   30 L   600 Cube feet content
D   1   4.47 Mean square of 5 × by 4

A simple case has been given, but the importance of finding the true mean square in cubical measurements must never be overlooked. (See Part 1, No. 8.)

When the cubical capacity is required in various measures, as cube yards, gallons, &c., and the dimensions are taken sometimes in feet, at others in inches, the proper gauge points, which are here given, must be used; if examined, they will be found to be the square

roots of the contents of the integer of measure used in the question
to be answered.

2. Dimensions being all in feet, and the answer in CUBE YARDS :

C Set the feet length   Find the cube yards contents
D To 5.196   Above the M. sq. feet of sides

3. Dimensions being all in feet, and the answer in GALLONS :

C Set the feet length   Find the gallons content
D To .4   Above the M. sq. feet of sides

4. Dimensions being all in feet, and the answer in BUSHELS :

C Set the feet length   Find the bushels contents
D To 1.131   Above the M. sq. feet of sides

5. Dimensions being all in inches, and the answer in CUBE FEET :

C Set the inches length   Find the cube feet content
D To 41.57   Above the M. sq. inches of sides

6. Dimensions being all in inches, and the answer in GALLONS :

C Set the inches length   Find the gallons contents
D To 16.65   Above the M. sq. inches in side

7. When the length is given in feet and the sides in INCHES :

C Set the feet length   Find the cube feet contents
D T. 12   Above the M. sq. in inches

Example of No. 7. A cistern is 40 inches long, its sides
30 × 25 inches, required its contents in gallons.

C 40 inches length   108.2 gallons contents
D 16.65   27.4 in M. sq. of 30 × 35

The Gauge point 16.65 = $\sqrt{277.27}$ cubic inches in a Gallon.

The cubing of unequal sided bodies may be effected as follows,
but the operation is less simple than the method given ; and is
inserted merely to compare the different workings ; the employment
of 4 lines of the Rule when 2 will effect the same purpose is always
objectionable, whatever facility practice may give in their use.

A   2nd   3rd { Then below the length on A
B Set one side of section on B   { Find cube ft. contents on B
(C Inverted) to other side ,,   on C
D   1st { Set the divisor*
                                   { To 10 on D

When Section	Length	*Divisors	
In^t. × In^t.	Inches.	1728	
In^t. × In^t.	Feet.	144	Answer
In^t. × F^t.	Feet.	12	in
F^t. × F^t.	F^t.		Cubic

Or any dimensions when the Section and } use 1 } Feet.
Length are in the same measure, and the }
answer of the same kind. }

8. CUBICAL PROPORTIONS.

A cubic foot contains—

1728 Cubic	inches	1.	Cube 1.	
2200 Cylindrical	,,	.7854	1.273 Cyl^r. 1.	
3300 Spherical	,,	.5236	1.909	2. Sphere 1
6600 Conical	,,	.2618	3.819	3. 2

And to reduce or compare Cubic with other Measures (on A and B).

A Set 55   Line of cubic   measures on A.
B To 70   ,,   cylindrical   ,,   on B.

A Set 42   ,,   cubic   ,,   on A.
B To 80   ,,   spherical   ,,   on B.

A Set 21   ,,   cubic   ,,   on A.
B To 80   ,,   conical   ,,   on B.

9. OF CYLINDRICAL FORMS (on C and D).

When the cube contents are required in the same measure as that in which the dimensions were given.

GENERAL RULE.

C Set the length   Find cube contents
D To 1.128   Above the mean diameter
                    c 2

Or, with circumference:

C Set the length   Find cube contents
D To 3.54   Above the circumference

Ex.: A cylinder is 6 feet long and 4 ft. 6 in. diameter, required its contents in cube feet.

C Set 6 feet   Find 95.4 cube feet contents
D To 1,128   Above 4.5 feet diameter

To Find the Capacity in Various Measures.

10. Dimensions being all in feet, and the answer in CUBE YARDS:

C Set feet length   Find the contents in cube yards
D To 5.86   Above mean diameter in feet

11. Dimensions being all in feet, and the answer in GALLONS:

C Set feet length   Find the contents in gallons
D To .451   Above mean diameter in feet

12. Dimensions being all in feet, and the answer in BUSHELS:

C Set feet length   Find the contents in bushels
D To 1.277   Above the mean diameter in feet

13. Dimensions being all in inches, and the answer in CUBE FEET:

C Set inches length   Find contents in cube feet
D To 46.9   Above mean diameter in inches

14. Dimensions being all in inches, and the answer in GALLONS:

C Set inches length   Find contents in gallons
D To 18.79   Above mean diameter in inches

Ex.: A cask being 40 inches long and 30 inches mean diameter, required the contents in gallons.

C   40 inches length   Find 102.3 gallons content
D *18.79   30 inches mean diameter

Note.—The gauge point * 18.79 is the square root of 353, the number of cylindrical inches in a gallon.

15. When the length is given in feet and diameter in INCHES:

C Set the feet length   Find the cube feet contents
D To 1.35   Above the M. diameter in inches

Or, with circumference:

C Set the feet length   Find the cube feet contents
D To 4.25   Above the circumference in inches

Note.—Cylindrical gauge points for Diameter $\times$ 3.14 give gauge points for circumference.

16. CYLINDRICAL PROPORTIONS.

A cylindrical foot contains—

1357.17 Cubic inches.	Cylinder	1.
2591.8 Spherical ,,	Sphere	2.
5183.6 Conical ,,	Cone	3.

17. To REDUCE AND COMPARE CYLINDRICAL WITH OTHER MEASURES (on A and B).

A Set 70   Line of cylindrical measures on A
B To 55   ,,   cubical measures on B

A Set 22   ,,   cylindrical measures on A
B To 33   ,,   spherical measures on B

A Set 22   ,,   cylindrical measures on A
B To 66   ,,   conical measures on B

18. CONTENTS OF CYLINDERS 1 FOOT IN DEPTH (C and D).

Diameters in feet.

C Set 42   Find cubic yards at 1 foot deep
D To 38   Above diameters in feet

C Set 7   Find cubic feet at 1 foot deep
D To 3   Above diameter in feet

C Set 490   Set gallons at 1 foot deep
D To 10   Above diameter in feet

19. Diameters in INCHES.

C Set 34   Weight of water in lbs. at 1 foot deep
D To 10   Above diameter of cylinder in inches

C Set 11   Cubic feet at 1 foot deep
D To 45   Above diameter of cylinder in inches

---

20. SPHERICAL FORMS (on C and D).

When the cube contents are required in the same measures as that
in which the dimensions were taken.

C Set the diameter   Find the cube contents
D To 1.382   Above the diameter

Or, for circumference:

C Set the diameter   Find the cube contents
D To 4.34   Above the circumference

Note.—The diameter of a sphere is taken twice, as it answers for
length.

Ex.: The diameter of a sphere being 4 feet, required its cube
contents.

C 4 diameter   33.5 cube feet content.
D 1.382   4 diameter.

When the cube contents are required in various measures, use the
prescribed gauge points.

21. Dimensions being all in feet, and the answer in CUBE YARDS:

C Set the feet diameter   Find the cube yards contents
D To 7.18   Above the diameter in feet

22. Dimensions being all in feet, and the answer in GALLONS:

C Set the feet diameter   Find the gallons contents
D To .553   Above the diameter in feet

23. Dimensions being all in feet, and the answer in BUSHELS:

C Set the feet diameter   Find the bushels contents
D To 1.563   Above the diameter in feet

24 Dimensions being all in inches, and the answer in CUBE FEET:

C Set the inches diameter   Find the cube feet contents
D To 57.44   Above the diameter in inches

25. Dimensions being all in inches, and the answer in GALLONS:

C Set the inches diameter   Find the gallons contents
D To 23.02   Above the diameter in inches

Example of No. 22. The diameter of a sphere being 3 feet 6 inches, required its contents in gallons.

C Set 3.5 diameter   140 gallons contents
D To .553   3.5 diameter.

26. SPHERICAL PROPORTIONS.

A Spherical foot contains
904.78 Cubical   inches
1152   Cylindrical ,,
3456   Conical   ,,

And to reduce or compare Spherical with other Measures (on A and B).

A Set 80   Line of spherical measures on A
B To 42   ,,   cubical   ,,   on B

A Set 33   ,,   spherical   ,,   on A
B To 22   ,,   cylindrical   ,,   on B

A Set 33   ,,   spherical   ,,   on A
B To 66   ,,   conical.   ,,   on B

27. SPHEROIDS (on C and D.)
Dimensions in same MEASURES.

C Set the fixed axis.   Find the cube contents
D To 1.302   Above the revolving axis

28. REGULAR CONES, RIGHT OR OBLIQUE (on C and D).

C Set the altitude *   Find the cube contents
D To 1.954   Above the diameter of base

Or, for circumference.

C Set the altitude   Find the cube contents
D To 6.14   Above the circumference of base

* Not the slant height.

29. FRUSTUMS OF REGULAR CONES (on A and B).

A { Set the sum of the diameters }   Find the cube contents
  { squared, less their product }
B   To 3.82   Above the altitude

30. CONES, REGULAR OR IRREGULAR (on A and B).

A Set the area of base   Find the cubic contents
B To 3   Above the altitude

31. FRUSTUM OF IRREGULAR CONES.

A { Set the sum of areas added }   Find the cube contents
  { to their mean proportional }
B   To 3   Above the altitude

32. PROPORTION OF CONES (on A and B).

A conical foot contains :

452.39 Cubic inches
576.   Cylindrical inches
864.   Spherical inches

To reduce or compare

A Set 80   Line of conical measures
B To 21   ,,   cubic   ,,

A Set 66   ,,   conical   ,,
B To 23   ,,   cylindrical ,,

A Set 66   ,,   conical   ,,
B To 33   ,,   spherical   ,,

33. SQUARE PYRAMIDS (on C and D).

C Set the altitude   Find the cube contents
D To   1·732   Above the side of base

34. FRUSTUM OF SQUARE PYRAMIDS (on A and B).

A Set { The *sum* of the sides squared less their product }   Find the cube contents

B To   3   Above the altitude

35. PYRAMIDS, REGULAR OR IRREGULAR (on A and B).

A Set the *area* of base   Find the cube contents
B To   3   Above the altitude

36. FRUSTUM OF REGULAR OR IRREGULAR PYRAMID (on A and B).

A Set { The *sum* of areas added to their mean proportional }   Find the cube contents

B To   3   Above the altitude

37. PARABOLOID (on A and B).

A Set the altitude   Find the cube contents
B To   2   Above the *area* of base

(On C and D).

C Set the altitude   Find the cube contents
D To   1·596   Above the *diamr.* of base

38. FRUSTUM OF PARABOLOID (on A and B).

A Set { The *sum* of the areas of the ends }   Find the cube contents

B To   2   Above the altitude.

39. CUBE CONTENTS OF REGULAR SOLIDS (on C and D).

No. of Facets.	Gauge Points.	
4	2·91	The length of the side, or
6	1·	edge of the facet given,
8	1·46	to find the cube con-
12	·36	tents
20	·698	

[See next page for Examples.]

C Set the length of edge   Find the cube contents
D To the G P   Above the length of edge

    Ex.: An eight-sided solid, with 6-inch edges, required the cube
    contents.

C   6-inch edge   103.3 inches cube contents
D   1·46 for 8 sides   6-inch edge

40. PRISMS, REGULAR (on C and D).

No. of Sides.	Guage Points.	
3	1.523	The length of the prism
4	1.	or shaft, and the mea-
5	.762	sure of the side of the
6	.62	section being given, to
7	.523	find the cube content
8	* .456	
9	.4	
10	.36	
11	.326	
12	.298	

C Set the length   Find the cube contents
D To the G P   Above the side

    Ex.: An octagonal prism being 10 inches long, and 4 inches in the
    side, required the cube contents.

C   10 inches   772·5 cube inches content
D   * ·456 G P for 8 sides   4   inches side

PART III.

MECHANICS.

1st. THE MECHANICAL POWERS.

Levers are of three orders—

The First having the Fulcrum between the Weight and Power.

„ Second „ Weight between the Fulcrum and Power.

„ Third „ Power between the Fulcrum and Weight.

Let F represent the fulcrum or centre of motion, P the power, W the weight or resistance.

LEVER OR WHEEL AND AXLE.

General Formula for any order of Lever.

A	Set the *distance* of P from F	Find W	or	Below W
B	To the *distance* of W from F	Above P		Find P

Ex.: The arms of a lever are 36 and 3 inches respectively, a power of 40 lbs. acts on the longer arm ; what weight will it balance ?

A	36 in. dis. of P	W 480	also	Ratio 12
B	3 in. dis. of W	P 40		Above 1.

Note.—The velocity ratio, or the motion of power and weight, whether in questions of simple or compound machines, is always shown on A above unity on B.

COMBINATION OF LEVERS OR WHEELWORK.

General Formula.

A	Product of dist. of powers	} from F	Find W	and ratio
B	„ „ weights		Above P	above 1.

Ex.: The product of the distance of powers from F is 300 inches, and the product of distance of weights is 20 inches in a compound machine of this class, with a power of 80 lbs. ; find weight and ratio of velocity.

A	15 ratio	First, Set {	300 in.	Find {	W 1200 lbs.
B	1		20 „		P 80 „

PULLEYS AND COMBINATION.
General Formula.

A Set the number of pulleys   Find W
B To   1   Above P

INCLINED PLANE AND WEDGE.

A Set the length   Find W (or resistance)   and r
B To altitude or breadth   Above P   1

> *Ex.:* What power will support 1000 lbs. weight on an inclined plane 14 feet long, sloping 5 feet high?

A (2·8 r)   Set   14 ft.   W 1000 lbs.
B (1)   To   5 ,,   P 357.15 answer

THE SCREW.
General Formula.

A Set the circumference described by P   Find W
B To the distance (centre) of threads   Above P

> *Ex.:* The circumference described by a power equal to 6 lbs. is 36 inches, and the threads of a screw ½ inch from centres of edge; find the weight.

A Set 36 in. cirfce.   W 432 lbs. r 72
B To   ·5 in. thread   P   6 ,,   1

COMPOUND MACHINES.
General Formula.

A Set the product of ratios of the several parts   Find W
B To   1   Above P   1

GENERAL FORMULA FOR ALL MECHANICAL POWERS.

A Set the motion of power   Find weight   and ratio
B To the motion of weight   Above power   1

Note.—Generally one-third more power than the results here found (W and P being in equilibrium) is required to overcome resistance owing to friction.

2ND. MACHINERY.

TOOTHED WHEELS (on A and B).

Note.—As the pitch of tooth, number of teeth, and diameter, are merely questions of simple proportion, and one of the terms, 3.14 = circumference of a circle whose diameter is 1, being constant, the operation in *all* cases is very easy, as the position of the terms in the statement are *fixed*, the conditions only varying.

General Rule.

A To 3.14 constant   Find N° of teeth   or   Below N° of teeth
B Set pitch of tooth then Above diam^r.   Find diam^r.

Or—

A Below 3.14   Set N°. of teeth
B Find pitch of tooth   To diam^r.

1. To find the number of teeth, pitch and diameter given.

Ex.: The pitch of tooth being 1.57 in., and diameter at pitch line 20 inches; required the number of teeth.

A To 3.14   40 No. of teeth answer
B Set 1.57   20 diam^r

2. To find diameter at pitch line, pitch and No. of teeth given.

Ex.: A pinion with 40 teeth, with 1.57 pitch, required the diameter at pitch line.

A To 3.14   40 teeth
B Set 1.57   20 diam^r. answer

3. To find the pitch of tooth, diameter at pitch line and N°. of teeth given.

Ex.: The diameter at the pitch line being 20 inches, and the N°. of teeth 40, required the pitch of tooth.

A Below 3.14   Set 40 teeth
B Find 1.57 pitch answer   To 20 in. diam^r.

Revolutions and Diameters.

1. To find the diameter of one wheel to any required number of tions in same time.

The revolutions of both wheels and } the diameter of the other wheel } being known.

(On A with slide *inverted*).

General Formula.

A   Find the ? reqd. diam^r.   Set ? the known diam^r.
S In^{,d} Above the ? reqd. revolt^{ns}. To ? the known revolt^{ns}.

 Ex.: One wheel 180 inches diam^r., making 9 revolutions per minute, another is reauired to make with it 30 revolutions, required its diameter.

A 54 required diam^r Set 18C
(S. In^r) 30 required revolutions   To 9

2. To find the diameters of two wheels to work at given velocities,

The distance between the centres } and number of required revolutions } being known.

(Direct proportions on A and B).

General Formula.

A Find ? Radius ? answer Set ? distance between centres
B Above ? *each* stated rev. ?   To ? the *sum* of revolutions

 Ex.: A shaft going 25 revolutions per minute, is to give motion to another of 15 revolutions in same time, the distance between centres 60 inches, required the diameters of the two wheels.

A Find 25 (required radii) and 42 Set 60 distance
B Above 15 (given rev.) and 25 To 40 *sum* of revolutions

3. To find the diameter required for a pulley to make more or less revolutions than one of stated diameter in equal time.

Revolutions and Diameter of one shaft } Revolutions only   of the other shaft } being known.

(On A with Slide inverted).

General Formula.

A Set the diam^r known of one shaft   Find   ? the required diameter
B To the revoltn^s   „   „   Above ? the given revolution
                                                        of other shaft.

Example :
1 Shaft making 50 revolutions, with a drum 30 inches diameter.
1 Shaft   „   40   „

required the diameter for a drum on the second shaft, to drive the
same *speed,* as 50 revolutions.

A   Set 30 inches dia.   Find   37.5 dia. for drum of 2nd shaft
(S. Int^r) To 50 revolutions   Above 40 revolutions

3RD. WEIGHT OF METALS.

THE full powers of the Slide Rule have never been fairly tested or
employed for ready and instant calculations ; the ordinary formula
is too much involved to make the operations apparent, as well as
easy.   It is only by availing ourselves of its proportional properties, that its capabilities stand in bold relief, either against
brain or book ; while one or both may be challenged for speed,
comprehensiveness, and accuracy; and for simplicity the following
method for finding the Weight of Metals cannot be approached.
The proportions are evident, and the operation instant and un-
hesitating; there is little to learn, and less to do, for the simple
setting of the constants gives off the widest range of tabular results
for each question of its class.   All ordinary forms of the various
metals in general use have been reduced to known proportions,
which can be tested by mere inspection ; the amount of time and
trouble saved is incredible to those who examine it.

The constants are formed for sectional areas, or diameters to one
foot in length, and give the weight in lbs., cwts., or tons.

The sides of square sided metal and ⎫ are given in eighths, inches,
    the diameter of cylindrical metal ⎭   and feet,
so that the weight of the smallest rod, or of the largest mass or
column, is arrived at with equal facility.

Use of the Constants explained.

Under the head of any metal, regard being had to its *form*, whether cubical or cylindrical, &c., will be found two constants showing the proportion of its area of section, or diameter, to weight in 1 foot lengths, which are to be set on the Rule, when the result is open to inspection; giving the weight of *all* sizes of that metal, the advantage is too evident to need urging.

Ex.: Required instantly, the weight per foot, of cast iron columns of *all* diameters.

Under *cylindrical* cast iron, inches in diameter, weight in lbs., we find

    C Set 10. *meaning* that 10lbs. is the weight of
    D To 2 inches diameter cylindrical iron.

Then with   Or, with weight in cwts.:

C Set 10   C is a line of lbs. weight   C Set 2.2 cwts.
D To 2   D   ,,   diameters   D To 10 inches diam.

and as C D are in proportion of squares and roots, it follows that the weight of *all* diameters are obtained by *one* setting.

There is no variation from this simple rule for metals in lengths · and the weight of metals by surface is equally easy.

---

1. CUBICAL METALS, ALL 1 FOOT LENGTHS (on A and B).

Note.—Find the area of the end or section in eighths, inches, or feet.

Ex.: Side × side = area of section in square units.

Description of Metal.		Lead.	Copper and Gun Metal.	Cast Brass.	Wrought Iron.	Cast Iron.	Cast Zinc.	Water* Compared	Result.
Find weight *in lbs.* ....on A	Set	2	3	2	4	1	2	·6	lbs.
Above any section *in 8ths*...on B	To	26	50	35	75	20	41	90	Sec.
Find weight *in lbs.* ....on A	Set	59	19	29	17	19	28	10	lbs.
Above any section *in in.* ...on B	To	12	5	8	5	6	9	23	Sec.
Find weight *in cwts.* .... on A	Set	4	2	1	3	2	1	1	Cwt.
Above any section *in in.* ...on B	To	91	59	31	100	71	36	260	Sec.
Find weight *in tons* ....on A	Set	7	10	7	5	10	10	2.5	Tons.
Above any section *in in.* ...on B	To	22	41	30	23	49	49	90	Sec.

* The specific gravity of *any other substance* being known, its weight can easily be found by comparison with the weight of water as given above, multiplying the weight of water thus obtained by the sp. gr. of the body of similar form and dimensions.

USEFUL DATA, showing Bases of the Constants, but not for use on the Rule.

For cubical metals.	Lead.	Copper and Gun Metal.	Brass.	Wrought Iron.	Cast Iron.	Cast Zinc.	Water.
Sp. gr. & oz. in cubic foot	11344	8788	8396	7788	7271	7190	1000
lbs. do.	709	549	524·75	486·75	454·44	449·37	62·5
cwts. do.	6·33	4·905	4·685	4·345	4·05	4·	·55
oz. in cubic inch	6·565	5·086	4·859	4·507	4·208	4·161	·579
lbs. do.	·4103	·3178	·3037	·292	·263	·26	·03617

2. CYLINDRICAL METALS, ALL 1 FOOT LENGTHS (on C and D).

			Lead.	Copper and Gun Metal.	Cast Brass.	Wrought Iron.	Cast Iron.	Cast Zinc.	Water.	Result.
Find weight in lbs. .... on C	Set		6	3	4·5	1·5	1·9	3·8	·26	lbs.
Above any diam. in 8ths on D	To		10	8	10	6	7	10	7	8ths.
Find weight in lbs. .... on C	Set		35	12	26	24	10	22	8·5	lbs.
Above any diam. in in. on D	To		3	2	3	3	2	3	5	in.
Find weight in cwts. ... on C	Set		3·4	6	2·55	9·5	2·2	5	6	cwts.
Above any diam. in in. on D	To		10	15	10	20	10	15	45	in.
Find weight in tons...... on C	Set		1	7	9	11	4	5·7	5	tons.
Above any diam. in ft. on D	To		2	6	7	5	5	6	15	ft

Ex. : A cylindrical column of cast iron is 7 inches inside diameter, and 8½ inches outside; required its weight per foot in cwts.

C   Set 2.2   Find 1.08 cwt.   Find 1.6 cwt.
D   To 10   Above 7 inches.   Above 8.5 inches.

1.6
1.08 deduct hollow.

———————

1.52 cwt. ⟹ 1.52 cwt. per foot.

Useful Data, and Bases of the Constants for Cylindrical Metals

	Lead.	Copper and Gun Metal.	Cast Brass.	Wrought Iron.	Cast Iron.	Cast Zinc.	Water.
Ounces in a cylindrical foot of	8909·6	6902	6594	6117	5711	5647	785·4
lbs. ,, ,,	557·	431·2	412·	382·3	357·	353·	49·1
cwts. ,, ,,	4·972	3·852	3·68	3·412	3·18	3·14	·432
Ounces ,, inch	5·156	3·99	3·815	3·54	3·305	3·268	·4547
lbs. ,, ,,	·322	·2496	·2385	·2215	·2066	·204	·02842

3. Spherical Metals, *any* diameter (on C and D).

		Lead.	Copper and Gun Metal.	Cast Brass.	Wrought Iron.	Cast Iron.	Cast Zinc.	Water.	Result.
Find weight *in ounces* on C	Set	Diam.	Diam.	Diam.	Diam.	Diam.	Diam.	Diam.	oz.
Above same dia. *in 8ths* on D	To	12·2	13·9	14·1	14·6	15·2	15·3	41·2	8ths.
Find weight *in lbs.* ... on C	Set	Diam.	Diam.	Diam.	Diam.	Diam.	Diam.	Diam.	lbs.
Above same dia. *in in.* on D	To	2·16	2·45	2·5	2·6	2·7	2·72	23·	in.
Find weight *in cwts*... on C	Set	Diam.	Diam.	Diam.	Diam.	Diam.	Diam.	Diam.	cwts.
Above same dia. *in in.* on D	To	22·8	26·0	26·6	27·6	28·6	28·8	77·	in.
Find weight *in tons* ... on C	Set	Diam.	Diam.	Diam.	Diam.	Diam.	Diam.	Diam.	tons.
Above same dia. *in ft.* on D	To	2·45	2·82	2·86	3·0	3·08	3·1	83·	ft.

Ex.: A spherical ball of solid copper is 8 inches in diameter, required its weight in lbs.

C   Set 8 inches diameter.   Find 85 lbs. answer.

D   To 2.45.   Above 8 inches diameter.

Bases of Constants. Spherical metals.	Lead.	Copper and Gun Metal.	Cast Brass.	Wrought Iron.	Cast Iron.	Cast Zinc.	Water.
Ounces in a spherical foot	5940	4601	4396	4078	3807	3764	523·6
lbs. ,, ,,	371·2	287·5	274·75	254·9	239·	235·3	32·73
cwts. ,, ,,	3·314	2·57	2·453	2·275	2·121	2·1	·288
Ounces ,, inch	3·437	2·663	2·54	2·36	2·2	2·18	·3032
lbs. ,, ,,	·2148	·1664	·159	·1476	·1377	·136	·0189

BAR IRON, ALL 1 FOOT LENGTHS.

Note.—The weight of *all* metals *in lengths* could be found by the foregoing constants for cubical and cylindrical metal, but bar iron being an article of constant use, and requiring frequent reference, each kind is separately treated of.

1. FLAT BAR IRON.

One Foot long. From 1 sixteenth to 1 inch thick. Any ins. wide.

Sixteenths of inch thick.	$1\frac{1}{16}$	$1\frac{3}{16}$	$1\frac{5}{16}$	$1\frac{7}{16}$	$1\frac{9}{16}$	$1\frac{11}{16}$	$1\frac{13}{16}$	$1\frac{15}{16}$
Find lbs. wt. per ft. on A set	2·5	7·5	12·5	17·5	22·5	27·5	32·5	37·5
Above any ins. wide on B to	12	12	12	12	12	12	12	12

Eighths of inch thick	$\frac{1}{8}$	$\frac{1}{4}$	$\frac{3}{8}$	$\frac{1}{2}$	$\frac{5}{8}$	$\frac{3}{4}$	$\frac{7}{8}$ Inch.
Find lbs. wt. per ft. on A set	5	10	15	20	25	30	35	40
Above any ins. wide on B to	12	12	12	12	12	12	12	12

2. SQUARE BAR IRON.

Eighths Square.	*Inches Square.*
C set 12 Find lbs. wt. per foot	C set 54 Find lbs. wt. per foot
D to 15 Above eighths sq. of sides	D to 4 Above ins. sq.

3. ROUND BAR IRON.

Eighths in Diameter.	*Inches in Diameter.*
C set 4·2 Find lbs. wt. per foot	C set 24 Find lbs. wt. per foot
D to 10 Above any eighths diam.	D to 3 Above any ins. diam

To FIND THE WEIGHT IN LBS. (AVDFSE.) OF ANY NUMBER OF SQUARE FEET SUPER. OF VARIOUS METALS, *one sixteenth thick,* FROM WHICH THE WEIGHT AT ANY THICKNESS MAY BE FOUND.

		Wrought Iron.	Cast Iron.	Cast Copper.	Gun Metal.	Cast Brass.
Find weight in lbs.	on A set	25	23	29	29	27 lbs.
Above any No. of sq. feet super.	on B to	10	10	10	10	10 sq. ft.

		Cast Lead.	Cast Zinc.	Cast Tin.	Cast Silver.	Pure Gold.
Find weight in lbs.	on A set	37	23	24	34	62·7 lbs.
Above any No. of sq. feet super.	on B to	10	10	10	10	10 sq. ft.

The easy method just given may fail to attract some who are wea-ded by habit to the use of the old gauge points for obtaining the weight of metals; but even these are greatly modified in the follow-ing Table, where, instead of taking the divisor on A, necessitating the use of four lines A B C D, the *square root* of it is employed. Every operation can then be performed on two lines (C and D) only, with equal accuracy and far greater readiness.

WEIGHT OF BODIES OBTAINED BY MEASUREMENT.

$$\text{Weight} = \frac{\text{Mean side or diameter}^2 \times \text{length.}}{\text{* Gauge Point.}}$$

* G P (divisor) on D is derived from the units of measure in an integer weight; viz., 2.16 is the square root of 4.66, the number of cubic feet of malleable iron in a ton. (See G P for iron, cubic, tons.)

Select the proper G P, con-
sidering whether for cube, cylinder, or sphere.
      „   dimensions in feet or inches.
      „   weight required in tons, cwts., or lbs.
    Also for what kind of substance.

The Gauge Point having to be adapted to all these conditions, it is important to remember that the dimensions must be taken—
    All in feet .... for t. ns.
    Length in feet, and the ends or diameter in ins. for cwts.
    All in inches.......................... for lbs.

Gencral Formula.

C Set the length   Find the weight on C.
D To the proper G P   Above the m. square or diameter

The *mean square* of cubical forms must be used.
The mean diameter of cylindrical forms must be used.

Table for Weight of Bodies by Measurement.

	CUBIC.			CYLINDER.			SPHERE
	Tons.	Cwts.	lbs.	Tons.	Cwts.	lbs.	lbs.
Lead . . .	1·777	4·768	1·562	2·	5·38	1·76	2·159
Gun Metal .	2·01	5·39	1·77	2·268	6·1	1·99	2·44
Copper, cast .	2·02	5·42	1·78	2·28	6·12	1·9	2·45
Brass, „ .	2·14	5·74	1·88	2·41	6·48	2·12	2·6
Iron, mal. .	2·16	5·8	1·9	2·44	6·55	2·21	2·66
Iron, cast . .	2·232	5·99	1·96	2·52	6·75	2 22	2·72
Zinc, „ .	2·236	6·	1·96	2·523	6·77	2·21	2·71
Marble .	3·601	9·66	3·19	4·1	10·9	3·6	4·4
Portland .	3·742	10·	3·27	4·22	11·33	3·7	4·5
Chalk . .	3·571	9·58	3·45	4·03	10·8	3·89	4·77
Clay . .	4·233	11·36	3·72	4·78	13·35	4·2	5·14
Gravel . .	4·32	11·59	3·8	4·87	13·1	4·28	5·25
Sand . .	4·32	11 59	3·8	4·87	13·1	4·28	5·25
Bricks . .	4·123	11·06	3·85	4·65	12·48	4·34	5 32
Earth . .	4·54	12 2	3·96	5·12	13·74	4·47	5·48
Coal . .	5· 4	14·33	4·69	6·02	16·16	5·29	6·48
Water . . .	5·98	16·05	5·253	6·75	18·1	5·93	7·26
Oak seasoned	6·93	18·58	5·46	7·8	20·96	6·16	7·55
Ash . . .	6·52	17·49	5·76	7·36	19·74	6·5	7·96
Elm . . .	8·12	19·59	6·4	8·24	22·1	7·4	8·86
Memel Fir .	8·22	21·76	7·13	9·27	24·5	8·	9·85
Yellow Pine .	8·59	22·5	7·85	9·7	25·5	8·85	10·84
The dimensions must be given.	All in feet for tons.	Feet long inches side for cwts.	All in inches for lbs.	All in feet for tons.	Feet long inches side cwts.	All in inches for lbs.	All in inches for lbs.

(Marble through Coal bracketed: Average)

Example of Working.

A cylindrical hollow column of cast iron is 15 feet long; inside diameter, 9 inches; outside diameter, 10 inches; required its weight in cwts.

For the right G P, look under cylinder, under cwts. for cast iron, find 6.75.

C	Set	15. feet long	26.62	32.87 cwt.
D	To	6.75 G P.	9 in.	10 in.

[Answer on next page.

Note. — When *any* length
is once set to a G P, the weight
to *all* diameters for that length
is given.

32.87

Deduct hollow 26.62

Weight of column 6.25 cwt.

Data to form Gauge Points, if required, for the following Metals :—

	Specific Gravity.	Lbs. in a Cubic ft.	Oz. in a Cubic inch.
Platinum, pure	21500	1343·8	12·44
Gold, do.	19258	1203·6	11·14
Gold, hammered.........	19361	1210·7	11·26
Silver, pure	10474	654·6	6·06
Silver, hammered	10510	656·9	6·08
Mercury, fluid	13568	848	7·85
Tin, cast.............	7292	455·7	4·22
Aluminum ......average	2600	162·5	1·5

To find the Weight in Lbs. of any Number of Square Feet Super. of Sheet Iron, the Thickness agreeing with the Number on the Wire Gauge.

			Numbers on Wire Gauge.						
Thickness Wire Gauge			1	2	3	4	5	6	7
			$\frac{5}{16}$			$\frac{1}{4}$			
Find weight in lbs.	on A Set		125	120	110	100	90	80	75
Above any No. of square feet	on B To		10	10	10	10	10	10	10

			Numbers on Wire Gauge.						
Thickness Wire Gauge			8	9	10	11	12	13	14
						$\frac{1}{8}$			
Find weight in lbs.	on A Set		70	60	56·8	50	46·2	43·1	40
Above any No. of square feet	on B To		10	10	10	10	10	10	10

			Numbers on Wire Gauge.						
Thickness Wire Gauge			15	16	17	18	19	20	21	22
Find weight in lbs.	on A Set		39·5	30	25	21·8	19·3	16·2	15	13·7
Above any No. of square feet	on B To		10	10	10	10	10	10	10	10

Ex. : Required the weight in lbs. of 27 feet super. of No. 10 gauge sheet iron.

A   To 56.8 lbs.   Find   153.4 lbs.   (Weight in lbs.)
B   Set 10 sq. feet.   Above   27 sq. feet.   (or *any* super. feet.)

To find the Weight in Lbs. of Lead, Copper, and Wrought Iron Pipes, 1 Foot Long, and any Circumference in Inches, 32nds Thick.

							32nds.
Thickness in 32nds of an inch				1	2	3	4
Lead...........	Find weight in lbs.	on A	Set	2	4	6	8
	Above circumference in inches	on B	To	13	13	13	13
Copper.........	Find weight in lbs.	on A	Set	1·7	3·4	4	3·4
	Above circumference in inches	on B	To	14	14	11	7
Wrought Iron	Find weight in lbs.	on A	Set	2·5	2·5	5	5
	Above circumference in inches	on B	To	24	12	16	12
Thickness in 32nds of an inch				5	6	7	8 ¼ in.
Lead...........	Find weight in lbs.	on A	Set	10	12	14	16
	Above circumference in inches	on B	To	13	13	13	13
Copper.........	Find weight in lbs.	on A	Set	8·5	8	11	29
	Above circumference in inches	on B	To	14	11	13	30
Wrought Iron	Find weight in lbs.	on A	Set	14	10	8	10
	Above circumference in inches	on B	To	26	16	11	12

Ex.: Required the weight in lbs. of 1 foot of copper pipe, 6 32nds thick, 20 inches circumference.

A	To	8 (lbs. per foot long)	Find	14.5 lbs.
B	Set	11 (in. circumference.)	Above	20 in. circumference.

To find the Weight in Lbs. of Cast Iron Pipes, 1 Foot Long, any Circumference in Inches, from ⅛th to 1¼ Inch Thick.

		$\frac{1}{8}$	$\frac{1}{4}$	$\frac{3}{8}$	$\frac{1}{2}$	$\frac{5}{8}$	$\frac{3}{4}$	$\frac{7}{8}$	Inch.	$1\frac{1}{8}$	$1\frac{1}{4}$
Thickness		$\frac{1}{8}$	$\frac{1}{4}$	$\frac{3}{8}$	$\frac{1}{2}$	$\frac{5}{8}$	$\frac{3}{4}$	$\frac{7}{8}$	Inch.	$1\frac{1}{8}$	$1\frac{1}{4}$
Find weight in lbs.	on A Set	7	7	20	16	22	28	33	32	36	40
Above the mean circumfe. in ins.	on B To	18	9	17	10	12	12	12	10	10	10

Note.—Two flanges generally reckoned as 1 foot.

Ex.: Required the weight in lbs. of 1 foot long of cast iron pipe, ⅝ths thick, 30 inches mean circumference.

A	To	22	Constants	Find 55 lbs.
B	Set	12	for ⅝ thick	Above 30 mean circumference

ARMOUR PLATING, WEIGHT IN LBS. AND CWTS., OF ANY NUMBER
OF SUPER. SQUARE FEET, FROM 1 TO 10 INCHES THICK (on
A and B.)

RESULT IN LBS.

Inches thick			1	1½	2	2½	3	3½	4
Find weight in lbs. ...... on	A	Set	406	608	812	1014	1217	1420	1622
Above *any* sq. ft. super. on	B	To	10	10	10	10	10	10	10
Inches thick			4½	5	5½	6	6½	7	7½
Find weight in lbs. ...... on	A	Set	1825	2028	2230	2434	2637	2840	3042
Above *any* sq ft. super. on	B	To	10	10	10	10	10	10	10
Inches thick			8	8½	9	9½	10	12	
Find weight in lbs. ...... on	A	Set	3245	3448	3650	3854	4056	4868	
Above *any* sq. ft. super. on	B	To	10	10	10	10	10	10	

Ex.: Find weight in lbs. of armour plate, 4½ inches thick, 9 feet
by 3 = 27 square feet.

A   Set   1825 lbs.   Find   4927 lbs.

B   To   10   Above 27 square feet.

RESULT IN CWTS.

Inches thick			1	1½	2	2½	3	3½	4
Find weight in cwts. ... on	A	Set	3·62	5·43	7·25	9·05	10·86	12·67	14·5
Above *any* sq. ft. super. on	B	To	10	10	10	10	10	10	10
Inches thick			4½	5	5½	6	6½	7	7½
Find weight in cwts. ... on	A	Set	16·3	18·1	19·9	21·7	23·5	25·34	27·16
Above *any* sq. ft. super. on	B	To	10	10	10	10	10	10	10
Inches thick			8	8½	9	9½	10	12	
Find weight in cwts. ... on	A	Set	28·97	30·78	32·58	34·39	36·2	43·4	
Above *any* sq. ft. super. on	B	To	10	10	10	10	10	10	

Ex.: Find the weight in cwts. of armour plate, 4½ inches thick,
9×3 = 27 square feet.

A   Set   16.3 (cwt.)   Find   44 cwt.

B   To   10 feet square.   Above 27 super. square feet.

WITH *weight only given*, IN LBS. OF SPHERICAL CAST IRON, TO FIND
DIAMETER.

A   To 7·55   Find a product * the cube root of which = dia.

B   Set 1.   Above the lbs. weight given

* Find the cube root of this product by cube root on $\dfrac{C}{D}$

Ex.: Find the diameter of a cast iron ball 28.5 lbs. weight.

A   To 7.55   Find 216 $\sqrt[3]{}$ = 6 inches diameter of ball.

B   Set 1   Above 28.5 lbs. given.

STEAM ENGINES, &c

THE ordinary rules are for nominal horse power only, and do not give a fair approximation even to the effective work of an engine. The following useful Slide Rule formula is adapted from the arithmetical process given in Templeton.

CONDENSING ENGINES (on C and D).

First. To the pressure in lbs. per *circular inch* × ·8 = ?

Add 3.73.

Sum. .

Multiply by the speed in feet per minute ?

Product* ?

C Set the product * Find the effective horse power
D To 182   Above (any) diam. of cylinder in inches.

Ex. : The pressure of steam being 6 lbs. on the *circular* inch, the speed of the piston 180 feet per minute, and the diameter of the steam cylinder 40 inches, find the effective horse power.

lbs. 6 × .8 = 4.8, add 3.73 = 8.53

Speed 180 ×

Product* 1535

C Set * 1535 74.4 effective horse power, answer.
D 182   40 in. diameter.

D

HIGH PRESSURE ENGINES (on C and D).

First. From the pressure in lbs. per circular inch   P
     Deduct due to the work of the engine itself   4 lbs.

                      Effective steam   Remainder.

Multiply by the speed of the piston in ft. per minute   P

                                  Product.

C   * product.   Find the effective horse power
D   182   Above (any) diam. of cylinder in inches.

Ex.: The pressure of steam being 30 lbs. on the circular inch, the speed of the piston 180 feet per minute, and the diameter of the steam cylinder 20 inches, find the effective horse power.

         lbs. 30, less 4 = 26 × 180 = 4680 product

C   Set   4680   57 horse power, answer
D   To   182   20 inches diameter.

To find pressure in lbs. per inch, Or diam. of cyl. to cwts. lift req⁴.

(S. Invt⁴) Set cwts. lift.   Find effective press. lbs. per inch
D   To 14.2   Above diam. of cylinder in inches.

Surface of water in boilers required to supply steam to cylinders of any diam. in inches.

C   Set   13   Find square feet of water surface required in boiler
D   To   10   Above diam. of any steam cylinder in inches.

Pressure of steam on square and circular inches compared.

A   Set   70   Pressure of steam on square inches ⎫
B   To   55   Ditto   circular inches ⎬ compared.
                                              ⎭

A   Set   45   Find column of mercury supported inches
B   To   22   Above pressure of steam, lbs. per *square* inch.

A   Set   53   Find column of mercury supported inches
B   To   20   Above pressure of steam, lbs. per *circular* inch.

REGULATORS (on A and B).

When the velocity (revolutions per minute) of the governor is given, to find length of penaulums.

A   Find * ? = square root of length.   To   187.5
B   Above 1   1st.   Set the number of
                                             revolutions per minute.

* For the square of which (the length of pendulums required) look to C and D.

Ex. : The velocity of a governor being 40 revolutions per minute, find the length.

A   Find 4.69 squared 22 in. length.   To   187.5
B   1   Set 40 revolutions.

When the length of the pendulums is given, to find the revolutions.

A   Find No. of revolutions.   To 187.5
B   Above 1.   Set ? square root of given length.

Ex. : The length of the pendulums being 22 inches, find the revolutions.

A   Find 40 revolutions, answer.   187·5
B   1   4·69 sq. root of 22, found
                                            on C and D.

CONSUMPTION OF FUEL, BY TONS OR CWTS., FOR MILES OR HOURS RUN.

Answer in lbs. per mile or hour.

SUPPLY IN CWTS.

A   Below 1.12.   To the number of miles or hours
B   Find lbs. per mile   Set the cwts. consumed.

SUPPLY IN TONS.

A   Below 22.4.   To the number of miles or hours
B   Find lbs. per mile   Set the tons consumed.

PART IV.

APPLICATION OF THE SLIDE RULE TO BUILDERS' WORK; FOR CIVIL ENGINEERS, SURVEYORS, CONTRACTORS, &c.

REDUCTION OF BRICKWORK FROM SUPER. OF VARIOUS THICKNESSES TO STANDARD (1½) CUBE FEET, CUBE YARDS, &c. (on A and B.)

Brickwork of various thicknesses reduced to Standard and Cube Work, &c.

						Standard						
No. of bricks thick............		¼	1	1¼	1½	1¾	2	2¼	2½	2¾	3
Compare											
Feet super....... on A	Set	300	300	60	100	60	75	60	60	54	50
Feet standard... on B	To	100	200	50	100	70	100	90	100	100	100
Feet super....... on A	Set	132	66	53	80	38	33	29	45	24	22
Feet cube ...... on B	To	50	50	50	90	50	50	50	85	50	50
Feet super....... on A	Set	240	180	140	120	28	90	80	70	65	60
Yards cube ... .. on B	To	5	5	5	5	4	5	5	5	5	5
Feet super. ...... on A	Set	816	408	326	272	234	204	182	163	148	136
Rods standard... on B	To	1	1	1	1	1	1	1	1	1	1
Feet super. ...... on A	Set	20	20	40	20	26	20	20	20	20	20
No. of bricks...... on B	To	110	220	550	330	500	440	500	550	600	661

Work given in inches thick.........		4½	5	9	9½	10	11	12	13	14
Feet super................. on A	Set	50	42	70	37	35	45	35	65	50
Feet standard............ on B	To	16	15	45	25	25	35	30	60	50
Feet super................. on A	Set	40	60	80	38	60	55	50	55	60
Feet cube .... on B	To	15	25	60	30	50	50	50	60	70

Work given in inches thick.........		14½	15	16	17	18	18¼	19	20
Feet super................. on A	Set	39	30	35	33	35	30	37	35
Feet standard.......... .. on B	To	40	32	40	40	45	40	50	50
Feet super................. on A	Set	75	60	60	60	60	65	60	60
Feet cube .... on B	To	90	75	80	85	90	100	95	100

When any of the above proportions either for bricks or inches thick are once set, the lines $\frac{A}{B}$ form Tables for comparing and reducing all super. of that thickness.

Note:—As reference may not be made to other parts of the work unconnected with building operations, it is re-stated that nothing is required but to set the constants found under any given thickness and opposite the reduction required, viz., to standard, cube, &c.

General Formula.

A Set { ? The constants   A is then a line feet super
B   { ? For thickness   B   „   reduced work

Ex.: Reduce 30, 80, 120, 250 feet super. 2½ bricks thick,

1st, to 1½ bricks (standard).

A Below 30   80   120   250 of 2½ bricks   Set 60
B Find   50   133.3 200   416.6 of 1½ bricks   To 100

2nd, to cube feet

A Below 30   80   120   250 of 2½ bricks   Set 45
B Find 56.6 151.1. 226.6 472.2 of cube feet   To 85

Decimal values of any feet in a rod.

A Find the decimal parts   Below any decimal parts 1 To ⎫
          or   ⎬ 1st
B Above any No. of feet   Find the value in feet 272 Sct ⎭

TABLES OF USEFUL PROPORTIONS FOR BUILDERS.

A Set   8   Below £ price per rod 1½ work
B To   7   Find pence price per foot super.

A Set 23   Below shillings per cube yard
B To   13   Find £ cost per rod.

A Set 50   Below shillings per load of 50 feet
B To   12   Find pence per cube foot.

A   Set   60   Below cubic yards
B   To   23   Find 1000^s of bricks required.

A   Set   34   Below cubic yards
B   To   3   Find rods standard ($1\frac{1}{2}$).

A   Set   90   Below cubic feet
B   To   80   Find standard feet ($1\frac{1}{2}$).

A   Set   34   Below cube yards
B   To   39   Find tons of brickwork.

A   Set   70   Below cube feet
B   To   60   Find cwts. of brickwork.

A   Set   45   Below number of bricks (5lb.)
B   To   2   Find cwts.

A   Set   9   Below feet super.
B   To   36   Find No. of paving bricks flat 9 ✕ 4 in.

A   Set   9   Below feet super.
B   To   48   Find No. of paving bricks on edge 9 ✕ 3 in.

A   Set   8   Below £s cost per acre
B   To   12   Find pence cost per square rod.

A   Set   123   Below £s cost per acre
B   To   6   Find pence cost per square yard.

A   Set   30   Below shillings cost per square rod
B   To   9   Find pence cost per square yard.

A   Set   50   Below shillings cost per load 50 feet
B   To   12   Find pence cost per cube foot.

TIMBER MEASURING.

(For superficial measurement, see Mensuration of Areas.)

---

1. ROUND TIMBER.

Note.—Instead of finding the quarter girth, it is easier by the Slide Rule to use a gauge point answering to the *whole girth*, *i.e.*, the G P for quarter girth, 12 × 4 = 48, the G P for whole girth. Both methods are shown.

The customary measure gives only about four-fifths of the *true* contents of round timber, proportioning it to timber when hewn; gauge points for the true contents are given also.

Allowance for Bark.	Customary Measure.		True Contents.
	Gauge Points for Whole Girth.	Gauge Points for Quarter Girth.	Gauge Points for Whole Girth.
None	48 on D	12 on D	42.5 on D
1-eighth	54.8	13.7	48.6
1-tenth	53.34	13.33	47.23
1-twelfth	52.36	13.1	46.37

> *Ex.:* A felled tree is 40 feet long, its mean girth 80 inches, no allowance for bark.

C Set 40 feet length	Find 111·1 cubic feet.
D To 48 G P.*	Above 80 ins. *whole* girth.

Common Method.

C Set 40 feet length	Find 111·1 cubic feet.
D To 12	Above 20 ins. *quarter* girth.

True Contents.

C Set 40 feet length	Find 142 cubic feet.
D To 42·5	Above 80 ins. whole girth.

---

2. HEWN TIMBER.

The usual method of measuring hewn timber is given below; but being satisfied that the Slide Rule is never more advantageously employed, nor easier learned, than when the *proportions* of a question are shown in the working, I recommend the following form, which answers for timber of equal or unequal sides, and for any size.

A To 144   Below *any* square inches in section
B Set length in feet.   Find cube feet contents.

For this purpose, and other questions in cubed timber, the 144 is marked near the centre of A, which gives the whole range of the line, the reading of the numbers on A and B then being

A 10 20 to 100 (144) 200 &c. to 1000   For sections on A.
B 1 2   10   20   100   For length and cube
                                       contents on B.

The simplicity of the process is apparent, for when 144 square inches in the cross section of any sided timber gives 1 foot cube for each foot in length, any given length in feet being set to 144, a table of sections on A and cube contents on B is formed; every increase to the right and decrease to the left of 144, marking its own proportion in cube contents. For an easy example take three sections each to 30 feet in length, viz.: $9 \times 8 = 72$, $12 \times 12 = 144$, and $12 \times 24 = 288$ inches.

A Below 72   To 144   Below 288
B Find 15 cubic feet   Set 30 feet length.   Find 60 cubic feet

Find 30 feet cube.

And so of any section, to any length set to 144, it being only necessary (before setting the length) to square up the inches in sides by common multiplication on A and B, as $20 \times 25 = 500$ sq. ins. in the cross section.

A To 20   Find 500 sq. inches in section
B Set 1   Above 25.

Note.—Smaller wood is sorted and sized for fuel into—
     Shids   4 feet long and 16, 23, 28, 33, and 38 inches girth
     Billets   3   ,,   7, 10, 14   ,,
     Cord wood, stacked, $3 \times 3 \times 14.22$ 128 c. ft. = 1 cord.
       Ditto   measuring $4 \times 4 \times$ 8 = 128   = ,,
     Billets   100   = ,,
     Ditto,   cwt. 10   = ,,
     Bavins and spray   100   = 1 load.

COMMON METHOD FOR HEWN TIMBER.

C Set the length in feet   Find cube feet content
D To 12.   Above the *mean sq.* in inches.

To find the mean square, say of 20 × 30 inches.

C Set 20 the smaller side   Below 30 the greater side
D To 20 the like number.   Find 24.5 the mean square.

Note.—In hewn timber of unequal sides, by the common method, the mean square must be taken. For example, with a section 30 in. by 20 in., the *true* mean square is 24.5, and *not* 25, as is sometimes erroneously used. The result in cubic measurements is very different, for in 40 ft. length of this section the

True mean square, 24.5, gives 166.6 content,
Arithmetical mean of 25 gives 173.2 content,

and the greater the difference of the sides the greater the error.

———

SCANTLINGS.

The rule given for Hewn Timber answers for all Scantlings,
or

A is a table of feet run   To 144 (marked on *centre*)
B is a table of feet cube   Set the square inches in sections.

Ex. : Find the cube contents to feet run of scantling, 6 × 6 inches = 36 ins. section. This easy example is given, but all others follow the same rule.

A Below 20 feet run 30 40   To 144
B Find   5 ft. cube 7.5 10 &c.   Set 36 the sq. ins. in cross sec.

To find the price per foot run of any scantling in the proportion to
any price per foot cube.

A Below any sq. ins. in section   To 144
B Find pence per foot run.   Set price in *pence* per foot cube.

Ex. : Timber at 3s. 6d. = 42d. per foot cube, find the price per foot run of scantling, 2½ × 4 = 10, 3 × 4 = 12, and 3 × 5 = 15.

A   Below 10   12   15 ins. section   ⎰To 144
B   Find 2.9   3.5 4.4 pence per ft. run. 1st.⎱Set 42 pence per foot
                                               cube.

D 3

REDUCTION OF DEALS AND BOARDS TO ST. PETERSBURG
STANDARD, 120 deals of 12′ 11″ × 1½″.

Feet run of any section reduced to feet run St. Petersburg.

Section in inches			4×12	4×11	4×10	4×9	4×8
Find run in feet of any section...... on A Set			495	540	594	660	742·5
Above run in feet St. Petersburg ... on B To			1440	1440	1440	1440	1440

Section in inches			4×7	4×6	3×12	3×11	3×10
Find run in feet of any section...... on A Set			848	990	660	720	792
Above run in feet St. Petersburg ... on B To			1440	1440	1440	1440	1440

Section in inches			3×9	3×8	3×7	3×6
Find run in feet of any section...... on A Set			880	990	1131·4	1320
Above run in feet St. Petersburg ... on B To			1440	1440	1440	1440

Section in inches			2½×12	2½×11	2½×10	2½×9
Find run in feet of any section...... on A Set			792	864	950·4	1056
Above run in feet St. Petersburg ... on B To			1440	1440	1440	1440

Section in inches			2½×8	2½×7	2½×6½	2½×6
Find run in feet of any section...... on A Set			1118	1357·7	1462	1584
Above run in feet St. Petersburg ... on B To			1440	1440	1440	1440

Example: Reduce 2000 feet run of 3×9 to feet run of St. Petersburg.

A   Set   880   Below 2000 or *any* feet run of 3×9
B   To   1440   Find   3300 feet St. Petersburg.

Feet Run of any Section reduced to St. Petersburg Deals.

Section in inches			4×12	4×11	4×10	4×9	4×8
Below feet run of any section...... on A Set			165	90	495	110	310
Find No. of St. Petersburg deals on B To			40	20	100	20	50

Section in inches			4×7	4×6	3×12	3×11	3×10
Below feet run of any section...... on A Set			460	750	110	120	330
Find No. of St. Petersburg deals on B To			65	91	20	20	50

Section in inches			3×9	3×8	3×7	3×6
Below feet run of any section............ on A Set			220	330	330	330
Find No. of St. Petersburg deals ...... on B To			30	40	35	30

Feet Run of any Section reduced to St. Petersburg Deals.

Section in inches			$2\frac{1}{2}\times12$	$2\frac{1}{2}\times11$	$2\frac{1}{2}\times10$	$2\frac{1}{2}\times9$
Below feet run of any section........... on A	Set	330	360	230	266
Find No. of St. Petersburg deals ...... on B	To	50	50	29	30

Section in inches			$2\frac{1}{2}\times8$	$2\frac{1}{2}\times7$	$2\frac{1}{2}\times6\frac{1}{2}$	$2\frac{1}{2}\times6$
Below feet run of any section........... on A	Set	495	430	390	660
Find No. of St. Petersburg deals ...... on B	To	50	38	32	50

Ex. : How many St. Petersburg deals are there in 550 feet run of 4 × 9.

A Set 110 Below 550
B To 20 Find 100 St. Petersburg deals.

This table is limited to a few of the most useful sizes; but by a little ingenuity *any conceivable section* can be *instantly* reduced to St. Petersburg standard, by the following

GENERAL RULE.

Note.—Although four lines are employed, C and D are only *once set*, as a register. D line being only used to set 5.1, the constant to 10, which cannot be reached on A. Note how the operations follow, 1st, 2nd, and 3rd.

A 2nd. { To *any* cross section (in. × in.) on A.
B { Set 10 (mid. No.) on B. 3rd. { Below any No. of ft. run on B.
C { Find No. of St. Ptbg. Dls. on C. 1st. { Set 5.1 on C.
D { To 10 on D.

Ex. : Reduce 330 feet run of 3 ×10 to St. Petersburg standard. Find how many deals (worked as above).

A 2nd. { To 3×10 in. = 30 sq. in. in section.
B { Set 10 3rd. { 330 feet run.
C { 50 deals, answer. 1st. { 5.1
D { 10

Note.—The St. Petersburg standard consists of 120 deals 12 feet long 11′ × 1½ inches.

Section	=	16.5	Square inches.
Total length	=	1440.	Feet run.
Ditto	=	17280.	Inches run.
Solidity	=	285120.	Inches cube.
Ditto	=	165.	Feet cube.
1 Deal	=	1.375	,, ,,
1 Deal	=	2376	Inches cube.

PRICES OF DEALS, &c., of *all sizes* (120 of 12 feet), compared with St. Petersburg standard at *any given rate.*

         Below *any other section.*

A 2×6=12   1st. Set 16.5 as 3×8 = 24 3×12 = 36
B £14·54 {To (say £20 } Find 29·1 £43·63.
  £14 10s. 11d. {Or *any* other rate.} Equal to £29 2s. £43 1. s. 6d.

To find the price per load of 50 cube feet, agreeing with St. Petersburg standard, or 120 of 12 feet, at any given rate.

A Below *any* given rate   1st. Set 32
B Find £ per load of 50 cubic feet   To 9.7 = £9 14s.

To find instantly the number of feet run of quartering of *any* section, to a load of 50 cubic feet.

A Find feet run to a load To 144. Find feet run to a load.
B Above 50 on B.   1st. { Set ? in.×in. } Above 50 on B.
           { Of any section. }

Note.—It is simple but useful to observe that in 120 of 12 feet of any size the cube feet contents equal 10 times the section.

Ex. : 3 in.×12 in. = 36 = 360 cubic feet. 4×11 = 44 = 440 cubic feet.

AVERAGE DIMENSIONS FOR SCANTLINGS OF ORDINARY STRENGTH
(on C and D.)

	Gauge Points for proper Section to Length.		
	Deal.	Elm.	Oak.
Ceiling joists	245	220	194
Common joists	380	342	300
Rafter	333	300	265
Do. principal	600	540	475
Common beams	512	460	410
Purlin	725	650	575
Summer	1080	970	860

General Formula.

C Set *any* inches thick   Below the length in feet
D To G P.   Find inches depth required.

AVERAGE STRENGTH OF ROPES, CHAINS, RODS, ETC.
(ON C AND D).

ROPES.

C   Set 6   Find weight in lbs. per fathom.
D   To 5   Above circumference in *inches.*

C   Set 7   Find cwts. safe bearing.
D   To 2   Above circumference in *inches.*

C   Sct 5   Find tons safe bearing.
D   To 7   Above circumference in *inches.*

CHAINS.

C   Set 2   Find the lbs. weights per fathom.
D   To 3   Above diameter in *sixteenths.*

C   Set 12   Find cwts. safe bearing.
D   To 4   Above diameter in *eighths.*

To find breaking weight in tons:—

Linked.		*Studded.*	
C Set 4 Find tons br. weight		C Set 8 Find tons br. weight.	
D To 6 Above sixteenths dia.		D To 8 Above 16″ diam.	

To find proof strength :—

Linked.		*Studded.*	
C Set 5 Find tons proof		C Set 7 Find tons proof	
D To 10 Above 16″ diam.		D To 10 Above 16″ diam.	

IRON RODS.

Square	*Round*
C Set 20 Find cwts. safe bearing	C Set 15 Find cwts safe bearing
D To 3 Above *side* in eighths	D To 3 Above *diam.* in eighths

BEAMS (1 *inch thick*).

Gauge points for Iron.	Oak.	Pine.	Fir.
To be used on C. 7.6	1.25	1	.75

GENERAL FORMULA FOR BEAMS SUPPORTED BOTH ENDS.

First find square root of length given; C and D gives this by *inspection;* then

C   Set G P for kind   Find cwts. safe load for beams 1 *inch thick.*

D   To *square root* of feet length   Above inches depth of beam.

Ex.: What is a safe load for an iron beam 2 inches thick, 6 inches deep (16 feet long, square root 4).

C   Set 7.6 G P iron   17 cwt. for *each inch* thickness

D   To 4, square root of length   6 inches deep.

CONTENTS of cylinders at 1 *foot deep,* diameter *in feet.*

C   Set   42   Find cube yards at 1 foot deep

D   To   38   Above any diam. of cylinder in feet.

C   Set   7   Find cube feet at 1 foot deep

D   To   3   Above any diameter in feet.

C   Set 490   Find gallons at 1 foot deep

D   To   10   Above any diameter in feet.

CONTRACTORS' WAGES TABLE (on A and B,)

Showing the amount in Shillings and Pence for any number of hours employed, at from 1 to 12 pence per hour.

Rate per hour in pence ...			1	1½	2	2½	3	3½	4	4½
Find shillings' amount ... on	A	Set	2	3	4	5	6	7	8	9
Above No. of hours' work on	B	To	24	24	24	24	24	24	24	24
Rate per hour in pence ...			5	5½	6	6½	7	7½	8	8½
Find shillings' amount ... on	A	Set	10	11	12	13	14	15	16	17
Above No. of hours' work on	B	To	24	24	24	24	24	24	24	24
Rate per hour in pence ...			9	9½	10	10½	11	11½	12	
Find shillings' amount ... on	A	Set	18	19	20	21	22	23	24	
Above No. of hours' work on	B	To	24	24	24	24	24	24	24	

Ex.: Wages due 40 hours' work at 4½*d.*

   A   Set   9   Find 15*s.*   Find shillings due

   B   To 24   Above 40, and above *any* hours at 4½*d.*

It is evident that a large pay sheet at various rates and hours may be made up and checked in a few minutes with the aid of the Slide Rule.

MANUAL AND HORSE LABOUR (on A and B.)

Gauge Points or Constants for Manual Labour.	Work Required.	Gauge Points or Constants for Horses' Work.
4000	Cubic feet of water raised	25300
148	Cubic yards ,, ,,	935
2232	Cwts. weight ,,	14100
112	Tons weight ,,	705
74	Cubic yards of earth ,,	475

1 foot high per hour.

General Formula.

A To G P   Work due

B Set feet high to be raised   Hours' labour.

Ex.: How many cubic yards of earth, to be raised 16 feet, are due to 20 hours' manual labour.

A G P 74   92 cube yards, answer

B   16 feet   20 hours.

Ex.: How many hours' manual labour are required to **raise** 92 cube yards of earth 16 feet high?

A G P 74   92 cube yards

B   16 feet   20 hours, **answer.**

PART V.

COMMERCIAL ARITHMETIC.

EXCHANGE OF MONEY, AND CONVERSION OF ENGLISH AND FOREIGN WEIGHTS AND MEASURES, WITH ENGLISH AND FRENCH STANDARDS IN DETAIL.

Note.—The proportional properties of the Slide Rule render invaluable assistance in Commercial Arithmetic, the most opposite denominations being by it as easily convertible as the more simple ones, altogether obviating the tedious process of reduction, as well as the necessity for bulky tables. Through the aid of fixed proportions, *all* others *of that kind* are instantly constructed by merely setting the rule as directed; by this simple method *any* Foreign and English standards may be compared as easily as if all were printed in a tabular form, an impossibility with variable exchanges, while, with the Slide Rule, *any* rate may be assumed at pleasure. The Examples given are arranged to convey a certain amount of information as to the relative value. Two or three points only demand attention.

1st. The line A always represents Foreign money, weight, &c.

     „   B always represents English   „

2nd. The proportions given must be maintained throughout the scale in each operation.

Example, showing the use of the constants and the simplicity of conversion. See French Measures, page 69.

A Set 76	Line of centimetres	100	120	450 cent^n
		thus,		
B To 30	„ inches	39.3	47.2	177 inches

Any constants, with the denominations affixed, are meant to show that the whole lines, right and left, are to be read as numbers of that kind. *Example:*

A 3	9	Set 12 gulden	30	36 gulden
		1st ‖		
B 5	15	To 20 shillings	50	60 shillings

Nothing is required but to set the numbers and observe the *proportions* throughout; all the commercial tables that follow are formed on this principle, which has been dwelt on to prevent the necessity for further explanation.

Country.	Money of Account. Gold & Silver.	Average Current Value. s. d.	Comparison with the £ English. (*Any* Rate of Exchange may be set to the £ sterling in lieu of the following.)
AMERICA, U.S.	1 dollar 100 cents.	4 2	A Set 4.80 dollars B To 1 £ English
AUSTRIA	1 gulden 100 neukreuzers	2 0	A Set 10 gulden B To 1 £
BELGIUM	1 frank 100 cents.	0 9½	A Set 25 francs B To 1 £
BRITISH INDIA	1 rupee 16 annas	2 0	A Set 10 rupees B To 1 £
DENMARK	1 rigsdaler 96 skilling	2 3	A Set 9 B To 1 £
FRANCE	1 franc 100 centimes	0 9½	A Set 25 francs B To 1 £
GERMANY	1 gulden 60 kreutzers	1 8¼	A Set 11.50 gul. & kt B To 1 £
HOLLAND	1 gulden 100 cents.	1 8¼	A Set 11.85 gulden B To 1 £
ITALY	1 lira 100 centissimi	0 9½	A Set 25 lira B To 1 £
PORTUGAL	1 milreis 1000 reis	4 6	A Set 4.444 milreis B To 1 £

Country.	Money of Account. Gold and Silver.	Average Current Value. s. d.	Comparison with the £ English. (*Any* Rate of Exchange may be set to the £ sterling in lieu of the following.)
Prussia	11 thaler 30 silbergrochen 12 pfennige	3 0	A Set 6.20 thalers &S.G. B To 1 £
Russia	1 ruble 100 copecs	3 2	A Set 6.25 ruble B To 1 £
Spain	1 dollar	4 2	A Set 4.16 dollars & reals B To 1 £
Switzerland	1 franc 100 centimes	0 9½	A Set 25 francs B To 1 £
Turkey	1 piaster 40 paras	0 2.1	A Set 115 piasters B To 1 £

WEIGHTS AND MEASURES.

——

AMERICA, U.S.   A Set 1   lb. } Weights similar.
                       B To 1   lb. }
Foot = 12 inches English

                       A Set 12 gallons, U.S.
                       B To 10 gallons, English

AUSTRIA   A Set 50 pfund
                       B To 62 lbs.
Foot = 12.445 in. Eng.

                       A Set 50 viertels
                       B To 156 gallons

BELGIUM

Metre = 39.371 inches

The weights and measures of Belgium, France and Holland, are assimilated under the metrical system (see details under France), the nomenclature only differing, viz., Holland, the pond is the kilogramme and the kunn is the litre; Belgium, the livre is the kilogramme, and the litron is the litre.

CHINA

Impl. foot = 12.162 inches

A Set 100 catties or pounds
B To 133 pounds weight

A Set 10 taus
B To 12 gallons

DENMARK

Foot = 12.357 inches

A Set 50 punds
B To 55 lbs.

A Set 50 viertels
B To 85 gallons

GERMANY

A Set 50 pfund
B To 53 lbs.
     Varying in States.

A Set 50 viertels
B To 80 gallons

ITALY

Florence foot = 11.94 inches
 ,, Braccio = 22.978

A Set 50 rotollo
B To 52 lbs.

A Set 10 barile
B To 163 gallons

PORTUGAL

A Set 100 arratel or pound
B To 101 pounds

A Set 10 almude (Lisbon) 10 (Oporto)
B To 364 gallons   552

PRUSSIA   A   Set 50 pfund
                           B   To 51 lbs.
Foot = 12.257 inches
                           A   Set 10 eimer
                           B   To 151 gallons

RUSSIA   A   Set 50 pounds or funt
                           B   To 45 lbs.
Foot = 13.75 inches
                           A   Set 10 vedras
                           B   To 27 gallons

SPAIN   A   Set 50 libra
                           B   To 51 lbs.
Foot = 11.13 inches
                           A   Set 10 arobas or cantaros
                           B   To 35 gallons

SWEDEN AND NORWAY   A   Set 50 skalpund
                           B   To 48 lbs.
                           A   Set 50 kanne
                           B   To 29 gallons

SWITZERLAND   A   Set 50 pfund
    Berne   B   To 57 lbs.
Foot = 11.54 inches   Varying in different Cantons.
                           A   Set 10 eimer
                           B   To 92 gallons

TURKEY   A   Set 50 rotolo
                           B   To 63 lbs.
Pic = 26.8 inches
                           A   Set 100 almud
                           B   To 115 gallons

Any foreign denomination may be so converted; having its value
singly in English, multiply it say by 50, and having found its equivalent, set the *constants* on A Set 50 of *any* kind foreign
                              B To = ? of given kind English.

COMPARISON AND CONVERSION OF FRENCH (METRICAL) AND
ENGLISH STANDARDS IN DETAIL.

LENGTH.

Ex.: A Set ? compare French standards on A
B To ? with English standards on B

A Set 76 millimètres
B To 3 inches

A Set 76 centimètres
B To 30 inches

A Set 76 decimètres
B To 25 feet

A Set 55 metres (unit of length)
B To 60 yards

A Set 11 decamètres
B To 120.3 yards

A Set 11 hectomètres
B To 1203 yards

A Set 50 kilomètres
B To 31 miles

A Set 50 myriamètres
B To 310.7 miles

ote.—The metre is the unit of length, and = 39.3708 inches or
3.2809 feet.

SOLID.

A Set 17 decistères
B To 60 cube feet

A Set 23 stères (unit = cube metre = 35.314
B To 30 cube yards   cube feet.)

A Set 23 decastères
B To 300 cube yards

WEIGHT.

A Set 1000 milligrammes
B To 15.4 grains

A Set 100 centigrammes
B To 15.4 grains

A Set 10 decigrammes
B To 15.4 grains

A Set 100 grammes (unit) **15.434 grains troy**
B To 1544 grains

A Set 20 decagrammes
B To 7. ounces avdp.

A Set 50 decagrammes
B To 16.1 ounces troy

A Set 20 hectogrammes
B To 70.5 ounces avdp.

A Set 50 kilogrammes $=$ 2 lb. 3 oz. $4\frac{1}{2}$ dr. avdp.
B To 110.3 lbs. avdp. -

A Set 10 myriagrammes
B To 220.6 lbs. avdp.

A Set 50 quintals $=$ 1 cwt. 3 qrs. $24\frac{1}{2}$ lbs.
B To 98.4 cwts.

A Set 61 millierbar $=$ 9 tons. 16 cwt. 3 qrs. $12\frac{1}{2}$ lbs.
B To 60 tons

SOLID.

A Set 410 cube centimètres
B To 25 cube inches

A Set 85 cube decimètres
B To 3 cube feet

A Set 23 cube metres $=$ **35.314 cubic feet**
B To 30 cube yards

CAPACITY.

A Set 1000 millitres
B To   61 cubic inches

A Set 100 centilitres
B To   61 cubic inches

A Set   10 decilitres
B To   61 cubic inches

A Set   50 { litres, unit of liquid capacity.
            { a cubic decimètre = 61.02379 cubic ins.
B To   88 pints

A Set   5 decalitres
B To   11 gallons

A Set   2 hectolitres
B To   44 gallons

A Set   2 kilolitres
B To   440 gallons (1 kilolitre = 35.3147 cubic feet)

A Set   23 myrialitres
B To   301 cubic yards

SURFACE.

A Set   50 centiares
B To   60 square yards

A Set   10 ares (unit) { square decamètre
                        { = 119.5991 square yards
B To 1196 square yards

A Set   20 ares
B To   80 square rods

A Set   49 decares
B To   12 square acres

A Set   30 hectares
B To 74·2 acres

AVOIRDUPOIS.

A   Set   53   grammes
B   To   30   drachms

A   Set   85   grammes
B   To   3   ounces

A   Set   50   kilogrammes
B   To   110·3 pounds

TROY.

A   Set   26   grammes
B   To   400   grains

A   Set   31   grammes
B   To   20   dwts.

A   Set   311   grammes
B   To   10   ounces

A   Set   30   kilogrammes
B   To   80   lbs.

SYSTÈME USUEL.

A   Set   100 pieds   &   Set 10 toises
B   To   109 feet   To 66 feet

A   Set   10 livres, 500 grammes each
B   To   11 lbs. avdp.

A   Set   20 livres
B   To   27 lbs. troy

A   Set   8 lieues poste
B   To   5 miles

REDUCTION OF QUANTITIES AND PRICES (on A and B).

1 on.

A   Set   5   Below   £ cost per ton.
B   To   5   Find   Shillings per cwt.

A   Set 28   Below   £ cost per ton.
B   To   3   Find   Pence cost per lb.

	A	Set	28	Below	£ cost per cwt.
	B	To	5	Find	Shillings per lb.
	A	Set	7	Below	£ cost per cwt
	B	To	15	Find	Pence per lb.
	A	Set	28	Below	Shillings cost per cwt.
	B	To	3	Find	Pence per lb.
Cwt.	A	Set	4	Below	£ cost per cwt.
	B	To	10	Find	Shillings per stone 14 lbs
	A	Set	8	Below	Shillings cost per cwt.
	B	To	12	Find	Pence per stone 14 lbs.
	A	Set	7	Below	£ cost per cwt.
	B	To	10	Find	Shillings per stone 8 lbs.
	A	Set	7	Below	Shillings cost per cwt.
	B	To	6	Find	Pence per stone 8 lbs.
Lb.	A	Set	4	Below	£ cost per lb.
	B	To	5	Find	Shillings per oz.
	A	Set	20	Below	Shillings cost per lb.
	B	To	15	Find	Pence per oz.
core.	A	Set	20	Below	Shillings per score.
	B	To	12	Find	Pence each
r 100	A	Set	25	Below	Shillings per 100
	B	To	3	Find	Pence each
.stone	A	Set	7	Below	Shillings per stone 14 lbs.
14 lb.	B	To	6	Find	Pence per lb.
.stone	A	Set	4	Below	Shillings per stone 8 lbs,
8 lb.	B	To	6	Find	Pence per lb.
Gross,	A	Set	60	Below	Shillings per gross (144)
144.	B	To	5	Find	Pence each

E

Acre.	A	Set	8	Below	£ cost per acre.
	B	To	1	Find	Shillings per square rod.
	A	Set	4	Below	£ cost per acre.
	B	To	6	Find	Pence per square rod.
	A	Set	123	Below	£ cost per acre.
	B	To	6	Find	Pence per square yard.
Square rod.	A	Set	20	Below	Shillings cost per square rod.
	B	To	8	Find	Pence per square yard.
Load 40 bush.	A	Set	6	Below	£ cost per load 40 bushels.
	B	To	3	Find	Shillings per bushel.
Quarter	A	Set	40	Below	Shillings per quarter.
	B	To	5	Find	Shillings per bushel.
Sack 168 lbs.	A	Set	42	Below	Shillings per sack (168 lbs).
	B	To	36	Find	Pence per gallon (7 lbs.)
Sack 280 lbs.	A	Set	40	Below	Shillings per sack (280 lbs.)
	B	To	12	Find	Pence per gallon (7 lbs.)
Flour.	A	Set	45	Below	Shillings per sack.
	B	To	6	Find	Pence per 4 lb. loaf.
Flour.	A	Set	70	Below	lbs. of flour used.
	B	To	90	Find	lbs. of bread made.
Barrel 36 gals.	A	Set	12	Below	Shillings per barrel (36 gals.)
	B	To	1	Find	Pence per quart.
£ per year.	A	Set	55	Below	£ expended per year.
	B	To	3	Find	Shillings per day.
	A	Set	35	Below	£ per year.
	B	To	23	Find	Pence per day.
	A	Set	13	Below	£ per year.
	B	To	5	Find	Shillings per week.

TROY AND AVOIRDUPOISE REDUCED (on A and B).

					Grains.	
A	Set	35	Below	Troy dwts.	of 24	each.
B	To	31	Find	Avdp. drachms	27.344	,,
A	Set	55	Below	Troy ounces	480	,,
B	To	60	Find	Avdp. ounces	437.5	,,
A	Set	204	Below	Troy ounces	480	,,
B	To	14	Find	Avdp. lbs.	7000	,,
A	Set	175	Below	Troy lbs.	5760	,,
B	To	144	Below	Avdp. lbs.	7000	,,
A	Set	6	Below	Troy lbs.	5760	,,
B	To	79	Find	Avdp. oz.	437.5	,,

JEWELLERS' WEIGHTS.

A	Set	4	Below	Troy grains
B	To	5	Find	Carat grains

GOLD, SILVER, AND COINAGE (on A and B.)

A	Set	39	Below Mint price in pence for standard gold, 934.5 pence per oz.
B	To	20	Find number of grains weight
A	Set	20	Below shillings per oz. troy
B	To	12	Find pence price per dwt.
A	Set	7	Below shillings per oz. troy
B	To	2	Find the carats fine, &c.
A	Set	93	Bank price in £ for standard gold
B	To	24	Number of oz. troy

WEIGHT IN COINS.

A	Set	67	Dwts. troy	Set 9 oz.	Set 3 lbs. troy
B	To	13	No. of sovs.	To 35 sovs.	To 140 sovs.
A	Set	40	Dwts. troy	Set 2 Oz. troy	
B	To	11	Shillings	To 11 Shillings	
A	Set	1	Oz. avdp.	Set 1 lb. avdp.	
B	To	5	Bronze halfpence	To 80 Bronze halfpence.	

E 2

SUPER. MEASURE, *whole price* in Shillings found (on A and B).

Dimensions, feet long, and feet broad.

„ inches „ inches „

The following method of pricing supers. will be found very expeditious and useful.

Price per foot or yard	No. 1. Gauge Points		Price.	No. 2. Gauge Points		Price.	No. 3. Gauge Points
1	240·		1	12·		1	1728·
1½	160·		1½	8·		1½	1150·
2	120·		2	6·		2	864·
2½	96·		2½	4·8		2½	690·
3	80·		3	4·		3	576·
3½	68·4		3½	3·42		3½	492·
4	60·		4	3·		4	432·
4½	53·2		4½	2·66		4½	362·
5	48·		5	2·4		5	345·6
5½	43·6		5½	2·18		5½	315·
6	40·		6	2·		6	288·
6½	37·		6½	1·85		6½	266·
7	34·2		7	1·71		7	246·24
7½	32·		7½	1·6		7½	230·
8	30·		8	1·5		8	216·
8½	28·2		8½	1·41		8½	204·
9	26·6		9	1·33		9	192·
9½	25·2		9½	1·26		9½	182·
10	24·		10	1·2		10	172·8
10½	22·8		10½	1·14		10½	163·
11	21·8		11	1·09		11	156·
11½	20·8		11½	1·04		11½	150·
12	20·		12	1·		12	144·
15	16·		15	·8		15	115·2
18	13·2		18	·66		18	96·
21	11·42		21	·571		21	82·32
24	10·		24	·5		24	72·
30	8·		30	·4		30	57·6
36	6·66		36	·333		36	48·
42	5·72		42	·286		42	41·6

Table 1 side notes: Below the other dimension, same kind, Find the whole price in £. A. To the G P. B. Set one dimension.

Table 2 side notes: Below the other side in yds. or feet, same kind, Find whole price of super. in shillings at the price per square yard or foot. A. To G P. B. Set one side in yards or feet.

Table 3 side notes: Below the other side in inches Find whole price of super. in shillings at the price per square foot. Below the other side in inches. A. To G P. B. Set one side in inches.

Ex.: What will 20 × 40 cost at 7½d. per square yard?

	G P	
A	32	40
B	20	£25
		answer.

Ex.: What will 15 × 9 cost at 2½d. per foot?

	G P	
A	4·8	15
B	9	28s. 1½d.
		answer.

Ex.: What will 50 × 40 inches come to at 7½d. per square foot?

	G P	
A	230	50
B	40	8s. 8d.
		answer.

SIMPLE INTEREST FOR DAYS, WEEKS, MONTHS, AND YEARS,

At any rate per cent., and for any principal.

The most unwieldly Interest Tables published are limited, compared with the range of the Slide Rule in working such questions; as a check upon such calculations its assistance is invaluable, for although, owing to the minuteness of the divisions on so small an instrument, the lower fractional parts are sometimes difficult to read, the correctness of the work is at once tested without labour or a chance of error. Almost any rate may be obtained by doubling the *result* of some one of the following, as twice the answer of $3\frac{1}{2} = 7$ per cent., &c. Or by *halving* the gauge point for any rate given below, we find a new one for double that rate.

General Formula.

A   To the gauge point for rate   Below the principal
B   Set the given time   Find the interest (see examples).

Interest shown in }	Shillings	Shillings	Shillings	£	
Rate per cent.	Gauge Point for days. No. 1.	Gauge Point for weeks. No. 2.	Gauge Point for months. No. 3.	Gauge Point for years. No. 4.	Rate per cent.
2	910·	130·	30·	50·	2
2 $\frac{1}{4}$	730·	·104·	24·	40·	$\frac{1}{2}$
3	602·	86·5	20·	33·2	3
3 $\frac{1}{8}$	580·	83·	19·2	32·	$\frac{1}{8}$
3 $\frac{1}{4}$	561·	80·	18·5	30·7	$\frac{1}{4}$
3 $\frac{3}{8}$	540·	77·2	17·8	29·6	$\frac{3}{8}$
3 $\frac{1}{2}$	522·	74·2	17·1	28·	$\frac{1}{2}$
3 $\frac{5}{8}$	503·	72·	16·5	27·6	$\frac{5}{8}$
3 $\frac{3}{4}$	486·	69·2	16·	26·6	$\frac{3}{4}$
3 $\frac{7}{8}$	470·	67·5	15·5	25·6	$\frac{7}{8}$
4	455·	65·	15·	25·	4
4 $\frac{1}{8}$	442·	63·5	14·5	24·2	$\frac{1}{8}$
4 $\frac{1}{4}$	430·	61·1	14·1	23·4	$\frac{1}{4}$
4 $\frac{3}{8}$	417·	59·5	13·6	22·8	$\frac{3}{8}$
4 $\frac{1}{2}$	405·	57·7	13·3	22·2	$\frac{1}{2}$
4 $\frac{5}{8}$	395·	56·5	12·9	21·6	$\frac{5}{8}$
4 $\frac{3}{4}$	384·	55·5	12·6	21·	$\frac{3}{4}$

Interest shown in }	Shillings	Shillings	Shillings	£

Interest shown in	Shillings	Shillings	Shillings	£	
Rate per cent.	Gauge Point for days. No. 1.	Gauge Point for weeks. No. 2.	Gauge Point for months. No. 3.	Gauge Point for years. No. 4.	Rate per cent.
4 $\frac{7}{8}$	375·	53·2	12·2	20·46	$\frac{7}{8}$
5	365·	52·	12·	20·	
5 $\frac{1}{8}$	356·	51·	12·7	19·6	$\frac{1}{8}$
5 $\frac{1}{4}$	347·	42·4	11·4	19·	$\frac{1}{4}$
5 $\frac{3}{8}$	340·	48·7	11·2	18·7	$\frac{3}{8}$
5 $\frac{1}{2}$	332·	47·2	10·9	18·2	$\frac{1}{2}$
5 $\frac{5}{8}$	324·	46·4	10 6	17·8	$\frac{5}{8}$
5 $\frac{3}{4}$	317·	45·1	10·4	17·4	$\frac{3}{4}$
5 $\frac{7}{8}$	310·	44·2	10·2	17·1	$\frac{7}{8}$
6	304·	43·2	10·	16 6	6
6 $\frac{1}{8}$	298·	42·3	9·8	16·3	$\frac{1}{8}$
6 $\frac{1}{4}$	292·	41·5	9·6	16·	$\frac{1}{4}$
6 $\frac{3}{8}$	286·	40·8	9·4	15·7	$\frac{3}{8}$
6 $\frac{1}{2}$	280·	40·	9 22	15·4	$\frac{1}{2}$
6 $\frac{5}{8}$	276·	39·2	9·1	15·1	$\frac{5}{8}$
6 $\frac{3}{4}$	271·	38·5	8·9	14 8	$\frac{3}{4}$
6 $\frac{7}{8}$	265·	37·7	8·7	14·5	$\frac{7}{8}$
7	260·	37·	8·55	14·24	7
7 $\frac{1}{8}$	256·	36 5	8 42	14·	$\frac{1}{8}$
7 $\frac{1}{4}$	252·	35·8	8·3	13·8	$\frac{1}{4}$
7 $\frac{3}{8}$	247·	35·2	8·2	18 5	$\frac{3}{8}$
7 $\frac{1}{2}$	243·	34·6	8·	13·32	$\frac{1}{2}$
7 $\frac{5}{8}$	239·	34·0	7·9	13·	$\frac{5}{8}$
7 $\frac{3}{4}$	235·	33 5	7·7	12·8	$\frac{3}{4}$
7 $\frac{7}{8}$	231·	33·	7·6	12 7	$\frac{7}{8}$
8	227·	32 5	7 5	12·44	8
8 $\frac{1}{8}$	224·	32·4	7 4	12·3	$\frac{1}{8}$
8 $\frac{1}{4}$	222·	31·5	7·26	12·1	$\frac{1}{4}$
8 $\frac{3}{8}$	219·	31·	7 2	11 9	$\frac{3}{8}$
8 $\frac{1}{2}$	215·	30·5	7·06	11·76	$\frac{1}{2}$
8 $\frac{5}{8}$	212·	30·	6 95	11 6	$\frac{5}{8}$
8 $\frac{3}{4}$	208·	29·6	6 85	11·4	$\frac{3}{4}$
8 $\frac{7}{8}$	205·	29·3	6 8	11 3	$\frac{7}{8}$
9	203·	28·75	6 66	11·1	9
9 $\frac{1}{8}$	200·	28 5	6 6	10 95	$\frac{1}{8}$
9 $\frac{1}{4}$	198·	28·	6·5	10 8	$\frac{1}{4}$
9 $\frac{3}{8}$	195·	25 7	6 4	10 65	$\frac{3}{8}$
9 $\frac{1}{2}$	192·	27 3	6·32	10 5	$\frac{1}{2}$
9 $\frac{5}{8}$	190·	27	6 25	10 4	$\frac{5}{8}$
9 $\frac{3}{4}$	187·	26·6	6·16	10 23	$\frac{3}{4}$
9 $\frac{7}{8}$	185·	26·3	6 1	10 1	$\frac{7}{8}$
10	182·	26·	6·	10·	10
Interest shown in	Shillings	Shillings	Shillings	£	

Ex. (from Column 1): What is the interest on £18 for 55 days at $3\frac{7}{8}$ per cent.

A G P 470 }
B   55 } for $3\frac{7}{8}$

£18
5.6s. = 5s. $7\frac{1}{4}d.$

The advantage of the Slide Rule is that the interest on *all* sums is shown for the number of days once set to G P for any rate, thus: for the above rate

A Set 470   Below £30   £43   £60
B   55   Find 3.5s.   5s.   7s.

Ex. (from Column 2 for Weeks): What is the interest on £65 for 40 weeks at $3\frac{1}{2}$ per cent.

A   £65   Set 72 G P for $3\frac{1}{2}$ per cent.
B   36.11   To 40 weeks
        $1\frac{1}{2}$

Ex. (from Column 3 for Months): What is the interest on £55 for 9 months at $3\frac{1}{8}$ per cent.

A   Set 19.2 G P for $3\frac{1}{8}$ per cent.   £55
B   To 9 months   25.78
                                 $9\frac{1}{2}$

To those who adopt the Slide Rule for interest calculations it is recommended to affix to this page decimal tables of weeks, months and days, which can be formed on the Rule and copied easily, as

A Decimals on A   1 Set
B To days in a week on B   7 To

A Table showing the Number of Days, from any Day in the Month to the same Day in any other Month.

To	Jan.	Feb.	Mar.	April.	May.	June.	July.	Aug.	Sept.	Oct.	Nov.	Dec.
January.................	365	31	59	90	120	151	181	212	243	273	304	334
February	334	365	28	59	89	120	150	181	212	242	273	303
March....................	306	337	365	31	61	92	122	153	184	214	245	275
April	275	306	334	365	30	61	91	122	153	183	214	244
May	245	276	304	335	365	31	61	92	123	153	184	214
June	214	245	273	304	335	365	30	61	91	122	153	183
July	184	215	243	274	304	335	365	31	52	92	123	153
August	153	184	212	243	273	304	334	365	31	61	92	122
September...............	122	153	181	212	242	273	303	334	365	30	61	90
October	92	123	151	182	212	243	273	305	335	365	31	61
November	61	92	120	151	181	212	242	273	304	334	365	31
December	31	62	90	121	151	122	212	243	274	304	335	365

From

COMPOUND INTEREST.

Rate per cent.

Years.	3	4	4½	5	5½	6	Years.
1	1·0300	1·0400	1·0450	1·05	1·055	1·06	1
2	1·0609	1·0816	1·0921	1·1025	1·1131	1·1236	2
3	1·0927	1·1218	1·1411	1·1576	1·1742	1·191	3
4	1·1255	1·1698	1·1925	1·2115	1·2390	1·2625	4
5	1·1592	1·2166	1·2462	1·2763	1·307	1·3382	5
6	1·1910	1·2663	1·3023	1·3401	1·3788	1·4185	6
7	1·2298	1·3159	1·3609	1·4071	1·4547	1·5033	7
8	1·2668	1·3685	1·4221	1·4771	1·535	1·594	8
9	1·3047	1·4233	1·4861	1·5513	1·6191	1·6895	9
10	1·3439	1·4802	1·5530	1·6289	1·7082	1·791	10

(left margin, vertical: Constants)

BY SLIDE RULE.

Ex. 1st. What will £550 amount to at compound interest in 4 years at 5 per cent.

A   To 1.211   Find 668.5 £ amount required
B   Set 1   Above 550

If for any term beyond the table, say 24 years, take some year by which it is divisible, and follow the series on the Rule, thus:

Ex. 2nd. £500 for 24 years at 5 per cent.

A   1.477   £738.5 (8 ys.)   1091.4 (16 ys.)   1612.5 (24 ys. required)
B   1   500   738.5   1091.4

Ex. 3rd. What sum will amount to £432 in 8 years at 3 per cent. compound interest.

A   To 1.267 (3 per cent. 8 ys.)   Below £432
B   Set 1   Find £341 sum required

The limits of such a work will not permit the application of the rule to the wide range of financial operations, which would occupy a volume; enough has been shown to prove the value of its assistance to the accountant. For the conversion of stock, exchanges, annuities, continuous and terminable, insurance, &c. &c., the saving of labour will be appreciated, especially in the absence of tables, logarithms,

&c., as a few constants which may be pencilled on the back of the rule itself suffice for particular questions.

PROPORTIONAL PARTS (on A and B).

The great number of *pro rata* questions that occupy the attention of commercial men must render any expeditious method of solution welcome; the peculiar applicability of the Slide Rule to such operations as dividends on shares, profits on separate advances, &c. &c., can easily be tested, and by its use proportions of one or many parts can be *instantly* found without any arithmetical process.

Ex.: Divide £80 in proportions of 2, 3, 5 = 10.

A	Below	2	3	5 parts		Set 10
B	Find	16	24	40 £		To £80

Ex.: Find the dividends at 8 per cent. on shares costing £35, 45, 62.10, and 70.

A	Below £35	£45	£62.10	£70	Set 100
B	Find £2.3	3.6	5	5.6	To 8 per cent.

Ex.: The assets being £120 and the debts £140, 80, 75, 50 = 345, find the respective dividends.

A	Below £50	75	80	140 debts	Set £345
B	Find £17.25	26	27.75	49 divds.	To 120

Further examples either of the simplicity or expedition of instrumental proportions are unnecessary.

PART VI.

SCIENTIFIC READINGS BY SLIDE RULE.

CONVERSION OF THERMOMETRIC SCALES (on A and B).

1. REAUMUR TO FAHRENHEIT.

A   Set   3   Read Reaumur   on A   { *Note.* After complying with
B ¹ˢᵗ To   9   Read Fahrenheit on B   {   the conditions.

Above zero, Reaumur.   *Add* 14.2 to the **given** degrees.

Ex.: Convert 20° Reaumur to Fahrenheit.

A   Set   4   20 add 14.2 = 34.2
B   To   9   Answer 77° Fahrenheit

Under zero, Reaumur, and { Use the *difference* between the
less than 14.2° given   }   given degrees, and 14.2

Ex.: Convert $\overline{7.1}$ Reaumur to Fahrenheit

A   Set   4   14.2 less $\overline{7.1}$ 7.1 the difference
B   To   9   Answer 16° Fahrenheit

Under zero, Reaumur, and { *Deduct* 14.2 from the **given**
more than 14.2° given   }   degrees.

Ex.: Convert $\overline{16°}$ Reaumur to Fahrenheit.

A   Set   4 $\overline{16°}$ less 14.2 = 1.8
B   To   9   Answer $\overline{4°}$ Fahrenheit

2. Fahrenheit and Reaumur.

A	Set	4	Read Reaumur on A	*Note.* After complying with
B	To	9	Read Fahrenheit on B	the conditions.

Above zero, Fahrenheit { Find, and use the *difference* between the given degrees and 32

Ex.: Convert 50° Fahrenheit to Reaumur.

A	Set	4	Answer 8° Reaumur
B	To	9	50 less 32 = 18, the *difference*

Below zero, Fahrenheit. Add 32 to the given degrees

Ex.: Convert $\overline{10°}$ under 0 Fahrenheit to Reaumur.

A	Set	4	Answer $18\frac{6}{9}$° Reaumur
B	To	9	10 add 32 = 42, the *sum*

3. Centigrade and Fahrenheit.

Note.—The method is precisely as with Reaumur, the constants only differing.

A	1st.	Set	5	Read Centigrade on A	*Note.* After complying
B		To	9	Read Fahrenheit on B	with the conditions.

Above zero, Centigrade. Add $17\frac{7}{9}$ to given degrees

A	Set	5	Below given degrees? added to 17.7 (sum)
B	To	9	Find the degrees of Fahrenheit (?)

Under zero, Centigrade, and *less* than $17\frac{7}{9}$ } Use the *difference* between the given degrees and $17\frac{7}{9}$.

A	Set	5	Below the *difference* ($17\frac{7}{9}$, *less* the given degrees)
B	To	9	Find the degrees (*above* zero) Fahrenheit.

Under zero, Centigrade, and *more* than $17\frac{7}{9}$ } *Deduct* $17\frac{7}{9}$ from the given degrees.

A	Set	5	Below the given degrees? less $17\frac{7}{9}$?
B	To	9	Find the degrees (*below* zero) Fahrenheit?

4. FAHRENHEIT AND CENTIGRADE.

Note.—The method as with Fahrenheit and Reaumur, the constants only changed.

A Set 5 Read Centigrade on A ⎧ *Note.* After complying witc
B To 9 Read Fahrenheit on B ⎨   the conditions.

Above zero, Fahrenheit ⎧ Find, and use the *difference* between the
                         ⎨   given degrees and 32.

A Set 5 Find the degrees of Centigrade?
B To 9 Above the *difference* between the given degrees and 32.

Below zero, Fahrenheit   *Add* 32 to the given degrees

A Set 5 Find the degrees of Centigrade   ?
B To 9 Above the given degrees *added* to 32   (sum).

5. CENTIGRADE AND REAUMUR.

Compared scales of Centigrade and Reaumur may be at once formed complete, without the conditions necessary to convert either to Fahrenheit.

A Set 5   ⎧ Below degrees Centigrade on A ⎫ or *vice*
          Then ⎨
B To 4   ⎩ Read degrees Reaumur   on B ⎭ *versá.*

SPECIFIC GRAVITY (on A and B).

TO FIND THE SP. GR. OF ANY BODY.

A Set the weight lost in water   Below 1000· sp. gr. of water
B To the whole weight.   Find ?   sp. gr. of substance

Ex.: A cubic foot of cast iron weighing 454 lbs., loses 62.5 lbs. when weighed in water. Required its sp. gr.

A Set 62.5 lbs. loss in water   Below   1000
B To 454 „ whole weight   Find   7271 sp. gr.

Having the contents in *cubic feet,* and sp. gr. of substance, to find ne weight in lbs. avdp.

A Set 7271 sp. gr. cast iron.   Find   454 lbs. avdp.
B To ·016 constant   Above (say) 1 cubic foot.

Having the contents in *cubic inches.* and sp. gr. of substance, to find the weight in lbs.

A Set 7271 sp. gr. cast iron. Find 454 lbs.
B To 27·73 constant Above (say) 1728 cubic inches.

UNIFORM MOTION (on A and B.)

A Set 1 Below miles per minute
B To 88 Find feet per second.

A Set 15 Below miles per hour
B To 22 Find feet per second.

A Set 5 Below miles per hour
B To 440 Find feet per minute.

A Set 23 Below miles = 5280 feet
B To 20 Find knots = 6075·8 „

ACCELERATED MOTION.

1st.—To find the space in feet fallen through in any given number of seconds by any heavy body (on C and D).

Note.—The whole space fallen through, is as the square of the seconds occupied in falling, × by 16·1.

Ex.: A stone dropped from a cliff is observed to be four seconds falling. Required the height.

C Set 400 Find 257 feet   Find height in feet
                              and
D To 5 Above 4 seconds.   Above *any* seconds descent·

2nd.—To find the velocity acquired during any given second of a body's descent (on A and B).

Note.—The spaces passed through for *each* second are as any given second of the descent doubled, *less* 1 and × 16·1.

Ex. : Through what space in feet will a body fall during the seventh second of its descent.

7 + 6 = 13

A Set 16 Find 209 feet during the seventh second
B To 1 Above 13, double of 7, less 1.

3rd.—To find the velocity acquired at the end of the time due to the height from which a body falls.

Note.—The velocity acquired at the end of the time is direct as 1, 2, &c., and the seconds × 32·2.

Ex. : What velocity will a falling body acquire at the end of eight seconds?

A   Set   32   Find 257 feet velocity at end of that time.
B   To   1   Above 8 seconds.

====

PROPORTIONS OF SPHERES.

SURFACE.

C   Set   50   Find surface in square super
D   To   4   Above diameter of sphere.

SOLIDITY.

C   Set   Diam.   Find cubic contents
D   To   4.37   Above diameter of sphere.

SIDE OF EQUAL CUBE.

A   Set   4   Find side of equal cube
B   To   5   Above diameter of sphere.

LENGTH OF EQUAL CYLINDER.

A   Set   4   Find height of equal cylinder
B   To   6   Above diameter of sphere.

WEIGHT OF WATER IN LBS. AVDP. IN ANY SPHERE;

C   Set   Diam.   Find lbs. water contained
D   To   7.26   Above diameter of sphere in inches.

====

FORCE OF WIND ON PERPENDICULAR SURFACE.

C   Set   12.3   Find force in lbs. on each sq. ft.
D   To   50   Above miles per hour of current.

TRAVELLING OF SOUND.

A	Set	14	Below seconds after report
B	To	3	Find miles distant.
A	Set	8	Below seconds after report
B	To	3000	Find yards distant.

VIBRATIONS OF PENDULUMS.

C	Set	39.14	Find inches length
D	To vibrations per minute.		Above 60

or

S (Inv^d.)	39.14	} Set	{ Table of inches length on S Inv*.
D	60		{ „ vibrations per minute on D.

WATER.

SUPPLY OF WATER FROM RAINFALL (on A and B).

In Symon's Register of Rainfall for 1865, the following formula is given :—

$$\left.\begin{array}{l}\text{Gallons daily}\\ \text{supply}\end{array}\right\} = 40,000 \times \left.\begin{array}{l}\text{Area in}\\ \text{sq. miles}\end{array}\right\} \times \begin{array}{l}\text{Inches available}\\ \text{Annual rainfall.}\end{array}$$

By the Slide Rule, the supply from any amount of rainfall over any area is instantly tabled.

General Rule.

A	To the G P 2.5	Below *any* inches annual rainfall
B	Set the area in sq. miles.	Find a number, to which *add* 5 ciphers, showing the gallons daily supply.

Ex.: From an available area of 21 square miles, with an annual rainfall of 15 inches, find the supply in gallons daily.

A	Set	2·5	Below 15 in. rainfall
			add 5 ciphers
B	To	21 sq. miles	Find 12,600,000 gallons daily.

Or for Rainfall per Acre :

A	Set	8	Below any inches annual rainfall.
B	To	500	Find gallons daily per acre.

FORCE OF WATER.   (C and D).

When the velocity is given in feet per second.

C   Set   1   Find the force in lbs. per sq. ft. (deduct $\frac{1}{30}''$ part)
D   To   1   Above the velocity in feet per second.

Ex.: The velocity being 9 feet per second, required the force in lbs. on an area of 6 square feet.

C   Set   6 square feet.   4S6, less $\frac{1}{30}$ = 474 lbs.
D   To   1   9 feet.

Velocity given in miles per hour.

C   Set   2.1   Find the force in lbs. per square foot.
D   To   1   Above the velocity in miles per hour.

Ex.: Required the *whole* pressure in lbs. on an area of 6 square feet, velocity being 4 miles per hour.

C   Set   2.1   33·3 lbs. per foot × 6 = 199.8 lbs.
D   To   1   4

PRESSURE OF WATER.   (on A and B).

Vertical Pressure.

Lbs. per square foot.

A   Set   500   Find the pressure per square foot in lbs.
B   To   8   Above any depth in feet.

Any area in square feet, cwts. pressure, to depth in feet.

A   Set the depth in feet   Find cwts. pressure
B   To   1·8   Above any area in square feet.

Any area in square feet, *tons* pressure, to depth in feet.

A   Set   depth in feet   Find tons pressure
B   To   G ᴾ 36.   Above *any* area in square feet.

Ex.: The depth being 20 feet and area 100 square feet, find the number of tons pressure.

A   20   56 tons
B   G P 36   100 square feet area.

Lateral Pressure.

A Below ½ depth.

B Find tons lateral pressure   1st. $\begin{cases} \text{To} & 36 \\ \text{Set} & \text{? area of side, ft.} \times \text{ft.} \end{cases}$

OVERSHOT WATERWHEELS.

To find the horse power.

1st. Cube the *radius* of the wheel in feet.

2nd. Find the square root of cubed number.

3rd. Find the square feet area in cross section of the stream.

    (All very simple operations by the Slide Rule).

General Rule.

A Find the horse power.

B Above square feet area (No. 3). 1st. $\begin{cases} \text{Set the sq. root (No. 2).} \\ \text{To} \qquad 6.5 \qquad \text{G P} \end{cases}$

Ex.: An overshot wheel is 30 feet diam., section of stream 6 square feet, required the horse power.

1st. C Set 15 radius.   Find 3375 $\begin{cases} \text{square root of which} \\ \text{on D is 58.} \end{cases}$

      D To   1   Above 15

Then:

A   Find 53.6 horse power.   Set 58 sq. root (No. 2).

B   Above 6 square feet area.   To 6.5 G P

PUMPS.

The handle of a pump is 60 inches long, the bucket arm is 21.5 inches, and the hand stroke 13.5 inches, required the bucket stroke.

A Find   7.6 in. bucket stroke (*a*)   To 13.5 the hand stroke

B Above 21.5 the bucket arm.   Set 38.5 differ. of length.

With a six-inch barrel, working at 20 feet depth, required the weight of water and gallons lifted.

C Set 20 feet.   Find 24.5 gall. = 245 lbs.

D To   5.4 G P   Above 6 in. diameter.

Power required to lift (same Example).

A　　　Find　123 lbs. power req^d.　　Set　245 lbs. lift
S (Inv^d). Above　38.5 long arm.　　To　21.5 bucket arm.

With a handle 60 inches in length, required the place of the fulcrum for a stroke of 15 inches, the *power* making 25 inches stroke.

To 25 add 15 = 40, sum of strokes.

A　Find (21.5 ins. Ful.)　Find　38.5　　Set　60 the length.
B　Above 15　　　　　　　Above 25　　To　40 the sum of strokes.

————

PUMPING ENGINES (on C and D).

To find the required diameter of a steam cylinder (at 10 *lbs.* effective pressure per inch) for a pump of *any* diameter, and at *any* yards deep.

Note.—The constants given for each diameter of pump, when once set on the Rule, form tables of yards deep on C, and of diameter of steam cylinder on D, thus:

C　　　⎧The given constants⎫　Below *any* No. of yards deep on C
　　Set⎨　　　　　　　　　　⎬
D　　　⎩for any dia. of pump⎭　Find the req^d. dia. of steam cyl. on D.

Ex.: Required the diameter of a steam cylinder (at 10 lbs. effective pressure) to work a 12-in. pump, 40 yards deep.

C　Set　22⎫　　Below　40 yards deep.
　　　　　　⎬
D　To　20⎭　　Find　27 in. diam. of steam cylinder required.

Diameter of pump in inches ...			3	4	5	6	7	8	9
Below any No. of yards deep... on	C	Set	87	50	32	22	10	49	25
Find diam. of steam cylinder on	D	To	10	10	10	10	25	20	16
Diameter of pump in inches ...			10	11	12	13	14	15	16
Below any No. of yards deep... on	C	Set	20	25	22	19	25	20	49
Find diam. of steam cylinder on	D	To	16	20	20	20	25	24	40
Diameter of pump in inches ...			17	18	19	20	21	22	23
Below any No. of yards deep... on	C	Set	50	20	15	40	22	20	30
Find diam. of steam cylinder on	D	To	43	27	25	43	35	35	45
Diameter of pump in inches ...			24	25	26	27	28	29	30
Below any No. of yards deep.. on	C	Set	22	20	25	20	40	15	14
Find diam. of steam cylinder on	D	To	40	40	46	43	63	40	40

Note.—The above constants are all calculated to 10 *lbs. per inch* effective pressure; but when it is required to find at greater or less pressure than that, *invert* the Slide to D, and taking the example given above.

S (Inv^d). Set 10 lbs.   Below *any other* given pressure

D   To 27 diam. at 10 lbs. pressure. } Find the required diameter.

as

S (Inv^d). Set 10   Below (say) 7 lbs. pressure (same pump)

D   To 27*   Find 32.4 ins. req^d. diam. at 7 lbs.

WEIGHT AND VOLUME OF WATER (on A and B, and C and D.)

A   Set   19   Below cubic inches of water
B   To   10   Find   ounces troy.

A   Set 160   Below cubic inches of water
B   To   7   Find   lbs. troy.

A   Set   26   Below cubic inches of water
B   To   15   Find   oz. avdp.

A   Set 416   Below cubic inches of water
B   To   15   Find   lbs. avdp.

A   Set   97   Below cubic feet of water
B   To   54   Find   cwts.

A   Set   36   Below cubic feet of water
B   To   1   Find   tons.

A   Set   4   Below cubic yards of water
B   To   3   Find   tons.

A   Set   22   Below cylindrical inches of water
B   To   10   Find   oz. avdp.

A   Set 350   Below cylindrical inches of water
B   To   10   Find   lbs. avdp.

A Set 20   Below cylindrical feet of water
B To   13   Find   cwts.

A Set 46   Below cylindrical feet of water
B To   1   Find   tons.

A Set 30   Below spherical inches of water
B To   15   Find   oz. avdp.

A Set 160   Below spherical inches of water
B To   3   Find   lbs. avdp.

A Set 31   Below spherical feet of water
B To   9   Find   cwts.

C Set 34   Below lbs. weight of water, 1 foot deep
D To   10   Find   diameter of cylinder in inches.

C Set   8   Below cubic feet, at 1 foot deep
D To   38   Find   diameter of cylinder in inches.

C Set 6.5   Below gallons in 3 feet length
D To   8   Find   diameter of pipe in inches.

A Set 100   Below gallons
B To   16   Find   cubic feet.

CASK GAUGING (on C and D).

Casks are gauged as cylinders, to which they are approximated, by finding a mean between the head and bung diameters. There are several methods, but the following simple rule answers all practical purposes, viz.: To find the mean diameter.

Multiply the difference between the head and bung diameters by ·63, and add the product to the head diameter for a mean diameter, thus :—

General Rule.

C   Set the length in inches Find the gallons content
D   To * G P 18.79 ╷ Above the mean diameter, ins.

Ex.: A cask is 40.7 ins. length.

   33.6 bung diameter ⎫
   25.8 head diameter ⎭ Mean diameter 31.3 ins.

  Difference 8.8 ×.63 = 5.5, *add* to head dia. 31.3

C   Set 40.7   Find 111 gallons content
D   To 18.79 G P   Above 31·3 mean diamr.

BUILDERS' TONNAGE OF SHIPS (on C and D).

C   Set the length in feet.† Find tons
D   To 13.75   Above breadth of beam.

Ex.: A vessel is 300 feet long, 30 feet beam, required the builder⁹ tonnage.

C 300 feet.   1400 tons.
D 13.75   30 feet beam.

RATIO, SPEED, AND POWER OF STEAM-SHIPS
(on A and B).

Ratio.

A   Set   immersed section Find the ratio
B   To   the power.   Above the *cube* ‡ of the speed.

Speed.

A   Set   immersed section Below the ratio
B   To   the power.   Find a No., ∛ of which = speed.‡

Power.

A   Set   immersed section. Find the power
B   To   the ratio.   Above the speed.

---

* The square root of cylindrical inches in a gallon.
† Between perpendiculars.
‡ The cube and cube root can both be found on C and D. See rule 6, page 10.

AGRICULTURAL.

LAND MEASURE (on A and B).

A great amount of labour is saved in casting areas by the use of the Slide Rule, as it answers for *all dimensions*, whether taken in chains, yards, feet, perches, &c.

General Rule.

A   To Gauge Point (divisor)   Below the breadth
B   Set the length   Find the contents.

Ex.: Required the area when 8.50 chains by 4.70.

A   Below 4.70 breadth   1st { To   10 G P
B   Find   4 acres.   { Set   8.50 length.

Note.—In the following forms the Gauge Points on the *right hand* are for Triangles, the whole base and perpendicular being given; the Gauge Points or common divisors are doubled for the area of triangles.

MEASURE IN CHAINS AND LINKS, TO FIND ACRES.

			For Triangles.
A	Set G P 10	Below chains broad	(G P 20)
B	To chains long	Find acres content.	

MEASURE IN PERCHES, TO FIND ACRES.

| A | Set G P 160 | Below perches broad | (320) |
| B | To perches long | Find acres. | |

MEASURE IN YARDS, TO FIND ACRES.

| A | Set G P 4840 | Below yards broad | (9680) |
| B | To yards long | **Find acres.** | |

MEASURE IN YARDS, TO FIND PERCHES.

For Triangles.

A	Set G P 30.3	Below yards broad	(60.5).
B	To yards long	Find perches.	

MEASURE IN LINKS, TO FIND PERCHES.

A	Set G P 625	Below links broad	(1250)
B	To links long	Find perches.	

MEASURE IN FEET, TO FIND PERCHES.

A	Set G P 272.2	Below feet broad	(544.5`
B	To feet long	Find perches.	

MEASURE IN MILES LONG, YARDS WIDE, TO FIND ACRES.

A	Set 11	Below yards wide
B	To 4	Find acres per mile length.

A	Set 1	Below chains wide
B	To 8	Find acres per mile length.

Note.—By *inverting* the slide to A, a table of yards wide and yards long, equal to 1 acre area, is found.

A	Set 4	Table { Yards wide on A
S Invᵈ. To 1210		{ Yards long on S Invᵈ. } = 1 acre.

REDUCING SCALES.

A Set 1	Below chains and links	or	A Set 50	Below links
B To 66	Read feet.		B To 33	Read feet.

A Set 24	Below links	or	A Decimal parts of a link	Set 1
B To 190	Read inches.		B Value in inches.	To 7.92

DECIMAL VALUES OF RODS IN AN ACRE, AND *vice versa.*

A	Line of decimal parts, as	.25 dec.	1st { To 1
B	Value in rods	40 rods, &c.	{ Set 160 rods.

MALT GAUGING.

Instead of the old method of gauge points, requiring constant practice, fixed numbers are given, which make the proportions of each question *evident*, and they can be used without hesitation or chance of error. The contents are calculated to 10 *inches deep;* the result being multiplied by *one-tenth* of the actual depth in inches, the whole contents are instantly found.

Note.—Find the square inches in area of the cistern; if square, by multiplying length and breadth on A and B, then—

General Rule for SQUARE CISTERNS (on A and B).

A   Set   9 bushels   Find the bushels to *each* 10 ins. deep
B   To   2000 sq. inches area   Above *any* number of sq. *ins.* area.

Ex.: A square cistern of malt is 32 inches deep, its sides being 84 and 60 inches, find bushels.

$$84 \times 60 = 5040 \text{ inches area.}$$

A   To   9   Find   22.72 bushels, at 10 inches deep
B   Set   2000   Above 5040
                        and to multiply

A   Set   3.2 $\begin{cases} \text{One-tenth of} \\ \text{actual depth} \end{cases}$ Find   72.7 bushels, whole contents
B   To   1   Above 22.72

The detail of the simplest operation *seems* long, while three or four seconds suffice to complete the above work by the Slide Rule.

When the length and breadth are given in *feet* and the depth in inches—

A   Set 13 bushels at 10 in. deep   Find bushels to *each* 10 in. deep
B   To 20 square *feet* area   Above *any* square *feet* area.

Ex.: A cistern 32 inches deep, sides 7 × 5 feet = 35 sq. ft.

A   Set 13 bushels   Find 22.7 bushels × 3.2 = 72.7 bushels conts.
B   To 20   Above 35 feet area given.

ROUND CISTERNS (on C and D).

Diameter and depth in *inches.*

C Set 10 bushels   Find bushels to each 10 inches deep

D To 53 in. dia.   Above *any* diameter in *inches.*

Ex.: A cistern is 32 inches deep, and 60 inches diameter ; required contents.

C Set 10   Find 12.75 × 3.2   48 bushels

D To 53   Above 60 inches diameter.

With diameter in *feet,* and depth in *inches.*

C Set 25 bushels   To find bushels at 10 inches deep

D To 7 feet   Above *any* diameter in inches

WEIGHT OF HAY IN STACKS.

To find the weight of hay in a stack, the dimensions in feet must first be carefully taken, viz. :—

	Length.	Breadth.	*True* height.
Square-sided stacks.	Length.	Breadth.	*True* height.
Round do.	{ Diameter or breadth { half-way up, for mean.		do.

* For the true height, measuring as a solid, add one-third of the height from the eaves to the top, to that of the eaves from the ground, then

Length × breadth × true height = cube contents of square stack.

Ex.: A stack is 30 feet long,
       20 feet broad,
       10 feet to eaves   10
       9 feet eaves to top $3 = \frac{1}{3}$

          ft. 13, true height.

Long.   Broad.   True height.
30 × 20 × 13 = 7800 cubic feet or 289 cubic yards contents.

F

		Cubic yards.		Cubic yards.		Cubic feet.
New hay	Ton =	26.6	Cwt. =	1.33	or	36
Settled hay	,, =	20.0	,, =	1.		27
Old hay	,, =	17.7	,, =	.885		24

Having found the contents in cube yards or cube feet, by far the simplest method is to use a constant for proportion on A and B for the *weight*, viz., so many cubic feet or cubic yards of each kind equalling 10 tons; the rest is then merely a question of quantity and weight.

SQUARE STACKS.

New Hay.

A Set   10 tons   10 tons   Find the weight in tons
B To 7200 cubic ft. or 266.6 cubic yds.   Above the contents in cubic yards or cubic ft,

Settled Hay.

A Set *10 tons   10 tons   Find the weight in tons
B To 5400 cubic ft. or 200 cubic yards.   Above the contents in cubic yards or cubic ft.

Old Hay.

A Set   10 tons   10 tons   Find the weight in tons
B To 4800 cubic ft. or 177.7 cubic yds.   Above the contents in cubic yards or cubic ft.

Ex.: A square-sided stack of *settled* hay is 30 feet long, 20 feet broad, true height 13 feet; required tons weight.

$$30 \times 20 \times 13 = 7800 \text{ cubic feet.}$$

A
B 1st { Set   10 tons   Find   14.4 tons.
        { To   5400* cubic feet   Above 7800 cubic feet.

ROUND STACKS.

Instead of employing the old method of Gauge Points, which present difficulties to those unaccustomed to their use, it is far easier to consider each round stack as a cylinder of some fixed length. When a constant is obtained, answering for *all* diameters, 10 *feet length* or *height* is assumed here, and for any *different height* the weight can be instantly found; say 10 feet high gives 25 tons; true height 23 feet.

2.3 multiply by

———————

57.5 tons for 23 feet high

New Hay.

Constants.
C Set 15 tons   7 tons   Find tons weight *each* 10 *feet high*

         or

D To 37 ft. dia.   80 ft. circ.   Above *any* diam. or circ. in feet.

Settled Hay.

Constants.
C Set 18* tons   15 tons   Find tons weight *each* 10 *feet high*

         or

D To 35 ft. dia.   100 ft. circ.   Above *any* diam. or circ. in feet.

Old Hay.

Constants.
C Set 20 tons   8 tons   Find tons weight *each* 10 *feet high*

         or

D To 35 ft. dia.   70 ft. circ.   Above *any* diam. or circ. in feet.

Ex.: A stack of settled hay is 40 feet mean diameter, 27 feet true height; required weight in tons.

C Set 18   Find 23.3 tons *each* 10 *feet. high* × 2.7 = 63 tons whole
     *   weight
D To 35   Above 40 feet diameter.

* Constants for Settled Hay

WEIGHT OF LIVE CATTLE (on C and D).

Considering the carcase of an animal as a cylinder of flesh of an average gravity, a close approximation may be made to its saleable weight when killed. Rules called cattle gauges are used, but the following is perhaps the simplest method, being without chance of error. The condition of the animal must, of course, be taken into account; the following is for "good or fair;" experience alone can guide the allowance to be made.

IN STONES OF 8 LBS.

G P
C Set 15 stones of 8 lbs. to *each*   Find the weight in stones of 8 lbs.
           10 *inches* of length   for each 10 inches length

D To 80 inches* girth.   Above any girth in inches

Ex. : A bullock is 90 inches in girth, and 55 inches in length ; required its weight in stones of 8 lbs.

C To 15   Find 19 stones for each 10 inches length × 5.5 =
                                                        104 stones

D Set 80   Above 90 inches girth

IN STONES OF 14 LBS.

G P
C Set 4 stones of 14 lbs. to *each* 10   Find weight in stones 14 lbs.
            inches of length   for each 10 inches length
D To 55 inches girth   Above any girth in inches

Ex : A bullock is 90 inches in girth, and 55 inches in length ; required its weight in stones of 14 lbs.

C Set 4 stones of 14 lbs.   Find 10.8 stones 14 lbs. for each 10 ins.
                             length × 5.5 = 59.4 stones of 14 lbs.

D To 55 inches girth   Above 90 inches girth.

* This admits of easier multiplying by any length taken in inches than when the length is given in feet.

CONSTRUCTION.

THE Sliding Rule having no "*works in it*," and there being nothing about its mechanism to account for its unerring action, many are puzzled by its performance. A brief explanation of the principle of construction is given; brief, because those acquainted with logarithms need none, having, I dare say, divined its secret, as a musician would read a scale, at sight; while others will derive little information, until they have extended their inquiries beyond the limits afforded here.

I can only describe how logarithms are employed on it. Each space occupied by 1 to 10 (A B C are double, D single lines) is supposed to contain 1000 or 10,000 parts, irrespective of its length; and so many of these parts are allotted to each number as is repre-sented by the logarithm of that number, thus :—

Occupies a space equal to

No. 1	whose log. is	0			1	—	0
,, 2	,,	301	of the 1000 parts; between 1 and	2
,, 3	,,	477	,,	,,	1	,,	3
,, 4	,,	602	,,	,,	1	,,	4
,, 5	,,	698	,,	,,	1	,,	5
,, 6	,,	778	,,	,,	1	,,	6
,, 7	,,	845	,,	,,	1	,,	7
,, 8	,,	903	,,	,,	1	,,	8
,, 9	,,	954	,,	,,	1	,,	9
,, 10	,,	1000	,,	,,	1	,,	10

All divisions and subdivisions are logarithmic parts also, and ex-pressed by log. numbers; in fact, we use the space (occupied by so many parts) engraved on the Rule, instead of printed numbers in a book, which mean the same quantity. To ensure accuracy, a dividing engine is used to lay down the scale.

Without entering into the history of logarithms, we have some-thing, even in the few given above, by which we can test their uses on the Rule; for instance, take

Multiply these numbers { No. 2 whose log. is 301 } add these logs.
{ and 4 ,, 602 }

8 ,, 903 == log. of 8

Now by the instrument (multiplication case 1), it is evident that by projecting the 1 on the Slide B to 2 on the line A, thus

```
A    1              2        4        8
B    Slide    ☞     1        2        4
```

we have in reality *added* the 301 parts between 1 and 2, to the place *first occupied* by the 4, which already held 602 spaces, making 903, and reaching 8 on A, therefore multiplying 4 (and *all other numbers* on the slide) by 2, the motion of the Slide, adding to or subtracting from its logarithmic *place*, in relation to the fixed numbers on A. This is the key to the working.

$\left.\begin{array}{c} C \\ D \end{array}\right\}$ form lines of squares and roots,

because the *space* allotted between each number on D, and *not* the *number* of parts, is exactly double that on C.

Were C made a triple line to D, cubes and cube roots, instead of squares and square roots, would be found by inspection; and other powers and roots could be so represented.

There are various kinds of Sliding Rules adapted to different uses; the principle is the same in all, the arrangements only differing to answer specific purposes; but the simple form here recommended will be found most generally useful, and its use most easily acquired.

I am greatly indebted to many writers on the instrument, whose able works scarcely reach the public, both for information and valuable suggestions. In naming one (Viscount Gage), the tribute is as much due to his earnest advocacy and recommendation of instrumental calculation, as to my grateful remembrance of his kindness.

CONCLUSION.

I CANNOT close this work without remarking, that there is hardly a limit to the use of the Slide Rule in ordinary questions of business. So far from the subject being exhausted, even by the wide range of examples given, it should be looked upon only as an introduction to instrumental calculation. Any person mastering its use will find endless opportunities for its employment, when in addition to its universal applicability in proportional statements, questions such as:

$$x = \frac{a^2 \cdot b}{c} \text{ or } x \frac{a_3 \cdot b}{c \sqrt{}}$$

are *instantly* solved, and are capable of being applied to a variety of operations, to the saving of time and unnecessary labour. In working with the Slide Rule, after a little practice, every step in the process appears as consequent as the figures in a sum, and as easily read by an intelligent eye; and the reason how, and *why* numbers are multiplied, and divided, squared and cubed by it, is as clear as in arithmetical solutions. It is unjust to call its operations merely mechanical, as far as the operator is concerned; for the ingenuity necessary to a correct method of stating each question is in itself a course of teaching, demanding mental exertion, but exempting it from the drudgery of the process.

The great value of the Slide Rule appears in its use in ordinary work; its assistance in gauging, timber measuring, mensuration of superficies and solids, gives only examples of its capabilities, which may be applied with equal advantage to the instant formation of tables of proportions, the comparison or conversion of things differing in quantity or value; in fact, it has been pronounced by those best capable of employing it, and estimating its powers, as "a marvel of ingenuity and utility." (*Guthridge.*)

One advantage to the public in the present work cannot fairly be overlooked: the publishers have decided on issuing with each copy a neat, portable, and perfectly reliable though inexpensive instrument.

answering all the purposes for calculating, of the article either costly or unwieldy, generally made. Attention to the examples given will illustrate its use; and as engineers, contractors, builders, architects, accountants, &c., have each their special subjects on which to employ it, so the practical mechanic will find in the work a store of valuable information, instructive and amusing. My chief aim has been to make the subject clear and its working easy. The application must rest with those who can appreciate a really useful instrument. It would be difficult to find one that so well repays the trouble *and pleasure* of investigation.

PRINTED BY J. S. VIRTUE AND CO., LIMITED, CITY ROAD, LONDON.

WEALE'S RUDIMENTARY SCIENTIFIC SERIES.

*** The volumes of this Series are freely Illustrated with Woodcuts, or otherwise, where requisite. Throughout the following List it must be understood that the books are bound in limp cloth, unless otherwise stated; *but the volumes marked with a ‡ may also be had strongly bound in cloth boards for 6d. extra.*

N.B.—In ordering from this List it is recommended, as a means of facilitating business and obviating error, to quote the numbers affixed to the volumes, as well as the titles and prices.

CIVIL ENGINEERING, SURVEYING, ETC.

No.

31. *WELLS AND WELL-SINKING.* By JOHN GEO. SWINDELL, A.R.I.B.A., and G. R. BURNELL, C.E. Revised Edition. With a New Appendix on the Qualities of Water. Illustrated. 2s.

35. *THE BLASTING AND QUARRYING OF STONE,* for Building and other Purposes. By Gen. Sir J. BURGOYNE, Bart. 1s. 6d.

43. *TUBULAR, AND OTHER IRON GIRDER BRIDGES,* particularly describing the Britannia and Conway Tubular Bridges. By G. DRYSDALE DEMPSEY, C.E. Fourth Edition. 2s.

44. *FOUNDATIONS AND CONCRETE WORKS,* with Practical Remarks on Footings, Sand, Concrete, Béton, Pile-driving, Caissons, and Cofferdams, &c. By E. DOBSON. Seventh Edition. 1s. 6d.

60. *LAND AND ENGINEERING SURVEYING.* By T. BAKER, C.E. Fifteenth Edition, revised by Professor J. R. YOUNG. 2s.‡

80*. *EMBANKING LANDS FROM THE SEA.* With examples and Particulars of actual Embankments, &c. By J. WIGGINS, F.G.S. 2s.

81. *WATER WORKS,* for the Supply of Cities and Towns. With a Description of the Principal Geological Formations of England as influencing Supplies of Water, &c. By S. HUGHES, C.E. New Edition. 4s.‡

118. *CIVIL ENGINEERING IN NORTH AMERICA,* a Sketch of. By DAVID STEVENSON, F.R.S.E., &c. Plates and Diagrams. 3s.

167. *IRON BRIDGES, GIRDERS, ROOFS, AND OTHER WORKS.* By FRANCIS CAMPIN, C.E. 2s. 6d.‡

197. *ROADS AND STREETS.* By H. LAW, C.E., revised and enlarged by D. K. CLARK, C.E., including pavements of Stone, Wood, Asphalte, &c. 4s. 6d.‡

203. *SANITARY WORK IN THE SMALLER TOWNS AND IN VILLAGES.* By C. SLAGG, A.M.I.C.E. Revised Edition. 3s.‡

212. *GAS-WORKS, THEIR CONSTRUCTION AND ARRANGEMENT;* and the Manufacture and Distribution of Coal Gas. Originally written by SAMUEL HUGHES, C.E. Re-written and enlarged by WILLIAM RICHARDS, C.E. Eighth Edition, with important additions. 5s. 6d.‡

213. *PIONEER ENGINEERING.* A Treatise on the Engineering Operations connected with the Settlement of Waste Lands in New Countries. By EDWARD DOBSON, Assoc. Inst. C.E. 4s. 6d.‡

216. *MATERIALS AND CONSTRUCTION;* A Theoretical and Practical Treatise on the Strains, Designing, and Erection of Works of Construction. By FRANCIS CAMPIN, C.E. Second Edition, revised. 3s.‡

219. *CIVIL ENGINEERING.* By HENRY LAW, M.Inst. C.E. Including HYDRAULIC ENGINEERING by GEO. R. BURNELL, M.Inst. C.E. Seventh Edition, revised, with large additions by D. KINNEAR CLARK, M.Inst. C.E. 6s. 6d., Cloth boards, 7s. 6d.

268. *THE DRAINAGE OF LANDS, TOWNS, & BUILDINGS.* By G. D. DEMPSEY, C.E. Revised, with large Additions on Recent Practice in Drainage Engineering, by D. KINNEAR CLARK, M.I.C.E. Second Edition, Corrected. 4s. 6d.‡

The ‡ indicates that these vols. may be had strongly bound at 6d. extra.

LONDON : CROSBY LOCKWOOD AND SON,

MECHANICAL ENGINEERING, ETC.

☞ *The ‡ indicates that these vols. may be had strongly bound at 6d. extra.*

MINING, METALLURGY, ETC.

4. *MINERALOGY*, Rudiments of; a concise View of the General Properties of Minerals. By A. RAMSAY, F.G.S., F.R.G.S., &c. Fourth Edition, revised and enlarged. Illustrated. 3s. 6d.‡

117. *SUBTERRANEOUS SURVEYING*, with and without the Magnetic Needle. By T. FENWICK and T. BAKER, C.E. Illustrated. 2s. 6d.‡

135. *ELECTRO-METALLURGY;* Practically Treated. By ALEXANDER WATT. Tenth Edition, enlarged, with additional Illustrations, and including the most recent Processes. 3s. 6d.‡

172. *MINING TOOLS*, Manual of. For the Use of Mine Managers, Agents, Students, &c. By WILLIAM MORGANS. 2s. 6d.'

172*. *MINING TOOLS, ATLAS* of Engravings to Illustrate the above, containing 235 Illustrations, drawn to Scale. 4to. 4s. 6d.

176. *METALLURGY OF IRON*. Containing History of Iron Manufacture, Methods of Assay, and Analyses of Iron Ores. Processes of Manufacture of Iron and Steel, &c.. By H. BAUERMAN, F.G.S. Sixth Edition, revised and enlarged. 5s.‡

180. *COAL AND COAL MINING*. By the late Sir WARINGTON W. SMYTH, M.A., F.R.S. Seventh Edition, revised. 3s. 6d.‡

195. *THE MINERAL SURVEYOR AND VALUER'S COMPLETE GUIDE*. By W. LINTERN, M.E. Third Edition, including Magnetic and Angular Surveying. With Four Plates. 3s. 6d.‡

214. *SLATE AND SLATE QUARRYING*, Scientific, Practical, and Commercial. By D. C. DAVIES, F.G.S., Mining Engineer, &c. 3s.‡

264. *A FIRST BOOK OF MINING AND QUARRYING*, with the Sciences connected there;;th, for Primary Schools and Self Instruction. By J. H. COLLINS, F.G.S. Second Edition, with additions. 1s. 6d.

ARCHITECTURE, BUILDING, ETC.

16. *ARCHITECTURE—ORDERS*—The Orders and their Æsthetic Principles. By W. H. LEEDS. Illustrated. 1s. 6d.

17. *ARCHITECTURE—STYLES*—The History and Description of the Styles of Architecture of Various Countries, from the Earliest to the Present Period. By T. TALBOT BURY, F.R.I.B.A.,;&c. Illustrated. 2s.
⁎ ORDERS AND STYLES OF ARCHITECTURE, *in One Vol.*, 3s. 6d.

18. *ARCHITECTURE—DESIGN*—The Principles of Design in Architecture, as deducible from Nature and exemplified in the Works of the Greek and Gothic Architects. By E. L. GARBETT, Architect. Illustrated. 2s.6d.
⁎⁎ *The three preceding Works, in One handsome Vol., half bound, entitled* "MODERN ARCHITECTURE," *price 6s.*;

22. *THE ART OF BUILDING*, Rudiments of. General Principles of Construction, Materials used in Building, Strength and Use of Materials, Working Drawings, Specifications, and Estimates. By E. DOBSON, 2s.‡

25. *MASONRY AND STONECUTTING:* Rudimentary Treatise on the Principles of Masonic Projection and their application to Construction. By EDWARD DOBSON, M.R.I.B.A., &c. 2s. 6d.‡

42. *COTTAGE BUILDING*. By C. BRUCE ALLEN, Architect. Eleventh Edition, revised and enlarged. With a Chapter on Economic Cottages for Allotments, by EDWARD E. ALLEN, C.E. 2s.

45. *LIMES, CEMENTS, MORTARS, CONCRETES, MASTICS,* ;PLASTERING, &c. By G. R. BURNELL, C.E Fourteenth Edition. 1s. 6d

57. *WARMING AND VENTILATION*. An Exposition of the General Principles as applied to Domestic and Public Buildings, Mines, Lighthouses, Ships, &c. By C. TOMLINSON, F.R.S., &c. Illustrated. 3s.

111. *ARCHES, PIERS, BUTTRESSES, &c.:* Experimental Essays on the Principles of Construction. By W. BLAND. Illustrated. 1s. 6d.

☞ *The* ‡ *indicates that these vols. may be had strongly bound at 6d. extra.*

LONDON : CROSBY LOCKWOOD AND SON,

116. *THE ACOUSTICS OF PUBLIC BUILDINGS;* or, The Principles of the Science of Sound applied to the purposes of the Architect and Builder. By T. ROGER SMITH, M.R.I.B.A., Architect. Illustrated. 1s. 6d.

127. *ARCHITECTURAL MODELLING IN PAPER*, the Art of. By T. A. RICHARDSON, Architect. Illustrated. 1s. 6d.

128. *VITRUVIUS — THE ARCHITECTURE OF MARCUS VITRUVIUS POLLO.* In Ten Books. Translated from the Latin by JOSEPH GWILT, F.S.A., F.R.A.S. With 23 Plates. 5s.

130. *GRECIAN ARCHITECTURE*, An Inquiry into the Principles of Beauty in; with an Historical View of the Rise and Progress of the Art in Greece. By the EARL OF ABERDEEN. 1s.

₊ *The two preceding Works in One handsome Vol., half bound, entitled "*ANCIENT ARCHITECTURE*," price 6s.*

132. *THE ERECTION OF DWELLING-HOUSES.* Illustrated by a Perspective View, Plans, Elevations, and Sections of a pair of Semi-detached Villas, with the Specification, Quantities, and Estimates, &c. By S. H. BROOKS. New Edition, with Plates. 2s. 6d.‡

156. *QUANTITIES & MEASUREMENTS* in Bricklayers', Masons', Plasterers', Plumbers', Painters', Paperhangers', Gilders', Smiths', Carpenters' and Joiners' Work. By A. C. BEATON, Surveyor. Ninth Edition. 1s. 6d.

175. *LOCKWOOD'S BUILDER'S PRICE BOOK FOR* 1895. A Comprehensive Handbook of the Latest Prices and Data for Builders, Architects, Engineers, and Contractors. Re-constructed, Re-written, and further Enlarged. By FRANCIS T. W. MILLER, A.R.I.B.A. 800 pages. 4s., cloth boards.

182. *CARPENTRY AND JOINERY*—THE ELEMENTARY PRINCIPLES OF CARPENTRY. Chiefly composed from the Standard Work of THOMAS TREDGOLD, C.E. With a TREATISE ON JOINERY by E. WYNDHAM TARN, M.A. Fifth Edition, Revised. 3s. 6d ƒ

182*. *CARPENTRY AND JOINERY. ATLAS* of 35 Plates to accompany the above. With Descriptive Letterpress. 4to 6s.

185. *THE COMPLETE MEASURER;* the Measurement of Boards, Glass, &c.; Unequal-sided, Square-sided, Octagonal-sided, Round Timber and Stone, and Standing Timber, &c. By RICHARD HORTON. Fifth Edition. 4s.; strongly bound in leather, 5s.

187. *HINTS TO YOUNG ARCHITECTS.* By G. WIGHTWICK. New Edition. By G. H. GUILLAUME. Illustrated. 3s. 6d.‡

188. *HOUSE PAINTING, GRAINING, MARBLING, AND SIGN WRITING:* with a Course of Elementary Drawing for House-Painters, Sign-Writers, &c., and a Collection of Useful Receipts. By ELLIS A. DAVIDSON. Sixth Edition. With Coloured Plates. 5s. cloth limp; 6s. cloth boards.

189. *THE RUDIMENTS OF PRACTICAL BRICKLAYING.* In Six Sections: General Principles; Arch Drawing, Cutting, and Setting; Pointing; Paving, Tiling, Materials; Slating and Plastering; Practical Geometry, Mensuration, &c. By ADAM HAMMOND. Eighth Edition. 1s. 6d.

191. *PLUMBING.* A Text-Book to the Practice of the Art or Craft of the Plumber. With Chapters upon House Drainage and Ventilation. Sixth Edition. With 380 Illustrations. By W. P. BUCHAN. 3s. 6d.‡

192. *THE TIMBER IMPORTER'S, TIMBER MERCHANT'S,* and BUILDER'S STANDARD GUIDE. By R. E. GRANDY. 2s.

206. *A BOOK ON BUILDING, Civil and Ecclesiastical,* including CHURCH RESTORATION. With the Theory of Domes and the Great Pyramid, &c. By Sir EDMUND BECKETT, Bart., LL.D., Q.C., F.R.A.S. 4s. 6d.‡

226. *THE JOINTS MADE AND USED BY BUILDERS* in the Construction of various kinds of Engineering and Architectural Works. By WYVILL J. CHRISTY, Architect. With upwards of 160 Engravings on Wood. 3s.‡

228. *THE CONSTRUCTION OF ROOFS OF WOOD AND IRON.* By E. WYNDHAM TARN, M.A., Architect. Third Edition, revised. 1s. 6d.

☞ *The ‡ indicates that these vols. may be had strongly bound at 6d. extra.*

Architecture, Building, etc., *continued.*

229. *ELEMENTARY DECORATION:* as applied to the Interior and Exterior Decoration of Dwelling-Houses, &c. By J. W. FACEY. 2s.

257. *PRACTICAL HOUSE DECORATION.* A Guide to the Art of Ornamental Painting. By JAMES W. FACEY. 2s. 6d.

*** *The two preceding Works, in One handsome Vol., half-bound, entitled* "HOUSE DECORATION, ELEMENTARY AND PRACTICAL," *price 5s.*

230. *A PRACTICAL TREATISE ON HANDRAILING.* Showing New and Simple Methods. By G. COLLINGS. Second Edition, Revised, including A TREATISE ON STAIRBUILDING. Plates. 2s. 6d.

247. *BUILDING ESTATES :* a Rudimentary Treatise on the Development, Sale, Purchase, and General Management of Building Land. By FOWLER MAITLAND, Surveyor. Second Edition, revised. 2s.

248. *PORTLAND CEMENT FOR USERS.* By HENRY FAIJA, Assoc. M. Inst. C.E. Third Edition, corrected. Illustrated. 2s.

252. *BRICKWORK :* a Practical Treatise, embodying the General and Higher Principles of Bricklaying, Cutting and Setting, &c. By F. WALKER. Third Edition, Revised and Enlarged. 1s. 6d.

23. *THE PRACTICAL BRICK AND TILE BOOK.* Comprising :
189. BRICK AND TILE MAKING, by E. DOBSON, A.I.C.E.; PRACTICAL BRICKLAY-
265. ING, by A HAMMOND; BRICKCUTTING AND SETTING, by A. HAMMOND. 534 pp. with 270 Illustrations. 6s. Strongly half-bound.

253. *THE TIMBER MERCHANT'S, SAW-MILLER'S, AND IMPORTER'S FREIGHT-BOOK AND ASSISTANT.* By WM. RICH-ARDSON. With Additions by M. POWIS BALE, A.M.Inst.C.E. 3s.‡

258. *CIRCULAR WORK IN CARPENTRY AND JOINERY.* A Practical Treatise on Circular Work of Single and Double Curvature. By GEORGE COLLINGS. Second Edition, 2s. 6d.

259. *GAS FITTING:* A Practical Handbook treating of every Description of Gas Laying and Fitting. By JOHN BLACK. 2s. 6d.‡

261. *SHORING AND ITS APPLICATION:* A Handbook for the Use of Students. By GEORGE H. BLAGROVE. 1s. 6d.

265. *THE ART OF PRACTICAL BRICK CUTTING & SETTING.* By ADAM HAMMOND. With 90 Engravings. 1s. 6d.

267. *THE SCIENCE OF BUILDING:* An Elementary Treatise on the Principles of Construction. By E. WYNDHAM TARN, M.A. Lond. Third Edition, Revised and Enlarged. 3s. 6d.‡

271. *VENTILATION:* a Text-book to the Practice of the Art of Ventilating Buildings By W. P. BUCHAN, R.P., Sanitary Engineer, Author of "Plumbing," &c. 3s. 6d.‡

272. *ROOF CARPENTRY;* Practical Lessons in the Framing of Wood Roofs. For the Use of Working Carpenters. By GEO. COLLINGS, Author of "Handrailing and Stairbuilding," &c. 2s.

273. *THE PRACTICAL PLASTERER:* A Compendium of Plain and Ornamental Plaster Work. By WILFRED KEMP. 2s. [*Just published.*

SHIPBUILDING, NAVIGATION, ETC.

51. *NAVAL ARCHITECTURE.* An Exposition of the Elementary Principles. By J. PEAKE. Fifth Edition, with Plates. 3s. 6d.‡

53*. *SHIPS FOR OCEAN & RIVER SERVICE,* Elementary and Practical Principles of the Construction of. By H. A. SOMMERFELDT. 1s. 6d.

53**. *AN ATLAS OF ENGRAVINGS* to Illustrate the above. Twelve large folding plates. Royal 4to, cloth. 7s. 6d.

54. *MASTING, MAST-MAKING, AND RIGGING OF SHIPS,* Also Tables of Spars, Rigging, Blocks; Chain, Wire, and Hemp Ropes &c., relative to every class of vessels. By ROBERT KIPPING, N.A. 2s.

☞ *The ‡ indicates that these vols. may be had strongly bound at 6d. extra.*

Shipbuilding, Navigation, Marine Engineering, etc., *cont.*

54*. *IRON SHIP-BUILDING.* With Practical Examples and Details. By Johh Grantham, C.E. Fifth Edition. 4s.

55. *THE SAILOR'S SEA BOOK :* a Rudimentary Treatise on Navigation. By James Greenwood, B.A. With numerous Woodcuts and Coloured Plates. New and enlarged edition. By W. H. Rosser. 2s. 6d.‡

80. *MARINE ENGINES AND STEAM VESSELS.* By Robert Murray, C.E. Eighth Edition, thoroughly Revised, with Additions by the Author and by George Carlisle, C.E. 4s. 6d. limp; 5s. cloth boards.

83bis. *THE FORMS OF SHIPS AND BOATS.* By W. Bland. Ninth Edition, with numerous Illustrations and Models. 1s. 6d.

99. *NAVIGATION AND NAUTICAL ASTRONOMY*, in Theory and Practice. By Prof. J. R. Young. New Edition. 2s. 6d.

106. *SHIPS' ANCHORS*, a Treatise on. By G. Cotsell, N.A. 1s. 6d.

149. *SAILS AND SAIL-MAKING.* With Draughting, and the Centre of Effort of the Sails; Weights and Sizes of Ropes; Masting, Rigging, and Sails of Steam Vessels, &c. 13th Edition. By R. Kipping, N.A., 2s. 6d.‡

155. *ENGINEER'S GUIDE TO THE ROYAL & MERCANTILE* NAVIES. By a Practical Engineer. Revised by D. F. M'Carthy. 3s.

55 & 204. *PRACTICAL NAVIGATION.* Consisting of The Sailor's Sea-Book. By James Greenwood and W. H. Rosser. Together with the requisite Mathematical and Nautical Tables for the Working of the Problems. By H. Law, C.E., and Prof. J. R. Young. 7s. Half-bound.

AGRICULTURE, GARDENING, ETC.

61*. *A COMPLETE READY RECKONER FOR THE ADMEA-* SUREMENT OF LAND, &c. By A. Arman. Fourth Edition, revised and extended by C. Norris, Surveyor, Valuer, &c. 2s.

131. *MILLER'S, CORN MERCHANT'S, AND FARMER'S* READY RECKONER. Second Edition, with a Price List of Modern Flour-Mill Machinery, by W. S. Hutton, C.E. 2s.

140. *SOILS, MANURES, AND CROPS.* (Vol. 1. Outlines of Modern Farming.) By R. Scott Burn. Woodcuts. 2s.

141. *FARMING & FARMING ECONOMY*, Notes, Historical and Practical, on. (Vol. 2. Outlines of Modern Farming.) By R. Scott Burn. 3s.

142. *STOCK ; CATTLE, SHEEP, AND HORSES.* (Vol. 3. Outlines of Modern Farming.) By R. Scott Burn. Woodcuts. 2s. 6d.

145. *DAIRY, PIGS, AND POULTRY*, Management of the. By R. Scott Burn. (Vol. 4. Outlines of Modern Farming.) 2s.

146. *UTILIZATION OF SEWAGE, IRRIGATION, AND* RECLAMATION OF WASTE LAND. (Vol. 5. Outlines of Modern Farming.) By R. Scott Burn. Woodcuts. 2s. 6d.

₊ *Nos.* 140-1-2-5-6, *in One Vol., handsomely half-bound, entitled* "Outlines of Modern Farming." By Robert Scott Burn. *Price* 12s.

177. *FRUIT TREES*, The Scientific and Profitable Culture of. From the French of Du Breuil. Revised by Geo. Glenny. 187 Woodcuts. 3s. 6d.

198. *SHEEP; THE HISTORY, STRUCTURE, ECONOMY, AND* DISEASES OF. By W. C. Spooner, M.R.V.C., &c. Fifth Edition, enlarged, including Specimens of New and Improved Breeds. 3s. 6d.‡

201. *KITCHEN GARDENING MADE EASY.* By George M. F. Glenny. Illustrated. 1s. 6d.‡

207. *OUTLINES OF FARM MANAGEMENT, and the Organi-* zation of Farm Labour. By R. Scott Burn. 2s. 6d.‡

208. *OUTLINES OF LANDED ESTATES MANAGEMENT.* By R. Scott Burn. 2s. 6d.

₊ *Nos.* 207 & 208 *in One Vol., handsomely half-bound, entitled* "Outlines of Landed Estates and Farm Management." By R. Scott Burn. *Price* 6s.

🖙 *The ‡ indicates that these vols. may be had strongly bound at 6d. extra.*

7, STATIONERS' HALL COURT, LUDGATE HILL, E.C.

Agriculture, Gardening, etc., *continued.*

209. *THE TREE PLANTER AND PLANT PROPAGATOR.* A Practical Manual on the Propagation of Forest Trees, Fruit Trees, Flowering Shrubs, Flowering Plants, &c. By SAMUEL WOOD. 2s.

210. *THE TREE PRUNER.* A Practical Manual on the Pruning of Fruit Trees, including also their Training and Renovation; also the Pruning of Shrubs, Climbers, and Flowering Plants. By SAMUEL WOOD. 1s. 6d.

*** Nos. 209 & 210 in One Vol., handsomely half-bound, entitled "*THE TREE PLANTER, PROPAGATOR, AND PRUNER.*" By* SAMUEL WOOD. *Price* 3s. 6d.

218. *THE HAY AND STRAW MEASURER:* Being New Tables for the Use of Auctioneers, Valuers, Farmers, Hay and Straw Dealers, &c. By JOHN STEELE. Fifth Edition. 2s.

222. *SUBURBAN FARMING.* The Laying-out and Cultivation of Farms, adapted to the Produce of Milk, Butter, and Cheese, Eggs, Poultry, and Pigs. By Prof. JOHN DONALDSON and R. SCOTT BURN. 3s. 6d.‡

231. *THE ART OF GRAFTING AND BUDDING.* By CHARLES BALTET. With Illustrations. 2s. 6d.‡

232. *COTTAGE GARDENING;* or, Flowers, Fruits, and Vegetables for Small Gardens. By E. HOBDAY. 1s. 6d.

233. *GARDEN RECEIPTS.* Edited by CHARLES W. QUIN. 1s. 6d.

234. *MARKET AND KITCHEN GARDENING.* By C. W. SHAW, late Editor of "Gardening Illustrated." 3s.‡

239. *DRAINING AND EMBANKING.* A Practical Treatise, embodying the most recent experience in the Application of Improved Methods. By JOHN SCOTT, late Professor of Agriculture and Rural Economy at the Royal Agricultural College, Cirencester. With 68 Illustrations. 1s. 6d.

240. *IRRIGATION AND WATER SUPPLY.* A Treatise on Water Meadows, Sewage Irrigation, and Warping; the Construction of Wells, Ponds, and Reservoirs, &c. By Prof. JOHN SCOTT. With 34 Illus. 1s. 6d.

241. *FARM ROADS, FENCES, AND GATES.* A Practical Treatise on the Roads, Tramways, and Waterways of the Farm; the Principles of Enclosures; and the different kinds of Fences, Gates, and Stiles. By Professor JOHN SCOTT. With 75 Illustrations. 1s. 6d.

242. *FARM BUILDINGS.* A Practical Treatise on the Buildings necessary for various kinds of Farms, their Arrangement and Construction, with Plans and Estimates. By Prof. JOHN SCOTT. With 105 Illus. 2s.

243. *BARN IMPLEMENTS AND MACHINES.* A Practical Treatise on the Application of Power to the Operations of Agriculture; and on various Machines used in the Threshing-barn, in the Stock-yard, and in the Dairy, &c. By Prof. J. SCOTT. With 123 Illustrations. 2s.

244. *FIELD IMPLEMENTS AND MACHINES.* A Practical Treatise on the Varieties now in use, with Principles and Details of Construction, their Points of Excellence, and Management. By Professor JOHN SCOTT. With 138 Illustrations. 2s.

245. *AGRICULTURAL SURVEYING.* A Practical Treatise on Land Surveying, Levelling, and Setting-out; and on Measuring and Estimating Quantities, Weights, and Values of Materials, Produce, Stock, &c. By Prof. JOHN SCOTT. With 62 Illustrations. 1s. 6d.

*** Nos. 239 to 245 in One Vol., handsomely half-bound, entitled "*THE COMPLETE TEXT-BOOK OF FARM ENGINEERING.*" By Professor* JOHN SCOTT. *Price* 12s.

250. *MEAT PRODUCTION.* A Manual for Producers, Distributors, &c. By JOHN EWART. 2s. 6d.‡

266. *BOOK-KEEPING FOR FARMERS & ESTATE OWNERS.* By J. M. WOODMAN, Chartered Accountant. 2s. 6d. cloth limp; 3s. 6d. cloth boards.

The ‡ indicates that these vols may be had strongly bound at 6d. extra.

LONDON : CROSBY LOCKWOOD AND SON,

MATHEMATICS, ARITHMETIC, ETC.

32. *MATHEMATICAL INSTRUMENTS*, a Treatise on; Their Construction, Adjustment, Testing, and Use concisely Explained. By J. F. HEATHER, M.A. Fourteenth Edition, revised, with additions, by A. T. WALMISLEY, M.I.C.E., Fellow of the Surveyors' Institution. Original Edition, in 1 vol., Illustrated. 2s.‡

** *In ordering the above, be careful to say, " Original Edition " (No. 32), to distinguish it from the Enlarged Edition in 3 vols. (Nos. 168-9-70.)*

76. *DESCRIPTIVE GEOMETRY*, an Elementary Treatise on; with a Theory of Shadows and of Perspective, extracted from the French of G. MONGE. To which is added, a description of the Principles and Practice of Isometrical Projection. By J. F. HEATHER, M.A. With 14 Plates. 2s.

178. *PRACTiCAL PLANE GEOMETRY:* giving the Simplest Modes of Constructing Figures contained in one Plane and Geometrical Construction of the Ground. By J. F. HEATHER, M.A. With 215 Woodcuts. 2s.

83. *COMMERuIAL BOOK-KEEPING.* With Commercial Phrases and Forms in English, French, Italian, and German. By JAMES HADDON, M.A., Arithmetical Master of King's College School, London. 1s. 6d.

84. *ARITHMETIC*, a Rudimentary Treatise on: with full Explanations of its Theoretical Principles, and numerous Examples for Practice. By Professor J. R. YOUNG. Twelfth Edition. 1s. 6d.

84*. A KEY to the above, containing Solutions in full to the Exercises, together with Comments, Explanations, and Improved Processes, for the Use of Teachers and Unassisted Learners. By J. R. YOUNG. 1s. 6d.

85. *EQUATIONAL ARITHMETIC*, applied to Questions of Interest, Annuities, Life Assurance, and General Commerce; with various Tables by which all Calculations may be greatly facilitated. By W. HIPSLEY. 2s.

86. *ALGEBRA*, the Elements of. By JAMES HADDON, M.A. With Appendix, containing miscellaneous Investigations, and a Collection of Problems in various parts of Algebra. 2s.

86*. A KEY AND COMPANION to the above Book, forming an extensive repository of Solved Examples and Problems in Illustration of the various Expedients necessary in Algebraical Operations. By J. R. YOUNG. 1s. 6d.

88. *EUCLID*, THE ELEMENTS OF : with many additional Propositions
89. and Explanatory Notes: to which is prefixed, an Introductory Essay 'on Logic. By HENRY LAW, C.E. 2s. 6d.‡

** *Sold also separately, viz. :—*

88. EUCLID, The First Three Books. By HENRY LAW, C.E. 1s. 6d.
89. EUCLID, Books 4, 5, 6, 11, 12. By HENRY LAW, C.E. 1s. 6d.

90. *ANALYiICAL GEOMETRY AND CONIC SECTIONS* By JAMES HANN. A New Edition, by Professor J. R. YOUNG. 2s.‡

91. *PLANE TRIGONOMETRY*, the Elements of. By JAMES HANN, formerly Mathematical Master of King's College, London. 1s. 6d.

92. *SPHERICAL TRIGONOMETRY*, the Elements of. By JAMES HANN. Revised by CHARLES H. DOWLING, C.E. 1s.
** *Or with " The Elements of Plane Trigonometry," in One Volume, 2s. 6d.*

93. *MENSURATION AND MEASURING.* With the Mensuration and Levelling of Land for the Purposes of Modern Engineering. By T. BAKER, C.E. New Edition by E. NUGENT, C.E. Illustrated. 1s. 6d.

101. *DIFFEREN₄IAL CALCULUS*, Elements of the. By W. S. B. WOOLHOUSE, F.R.A.S., &c. 1s. 6d.

102. *INTEGRAL CALCULUS*, Rudimentary Treatise on the. By HOMERSHAM COX, B.A. Illustrated. 1s.

136. *ARITHMETIC*, Rudimentary, for the Use of Schools and Self-Instruction. By JAMES HADDON, M.A. Revised by A. ARMAN. 1s. 6d.
137. A KEY TO HADDON'S RUDIMENTARY ARITHMETIC. By A. ARMAN. 1s. 6d.

☞ *The ‡ indicates that these vols. may be had strongly bound at 6d. extra.*

7, STATIONERS' HALL COURT, LUDGATE HILL, E.C.

Mathematics, Arithmetic, etc., *continued.*

168. *DRAWING AND MEASURING INSTRUMENTS.* Including—I. Instruments employed in Geometrical and Mechanical Drawing, and in the Construction, Copying, and Measurement of Maps and Plans. II. Instruments used for the purposes of Accurate Measurement, and for Arithmetical Computations. By J. F. HEATHER, M.A. Illustrated. 1s. 6d.

169. *OPTICAL INSTRUMENTS.* Including (more especially) Telescopes, Microscopes, and Apparatus for producing copies of Maps and Plans by Photography. By J. F. HEATHER, M.A. Illustrated. 1s. 6d.

170. *SURVEYING AND ASTRONOMICAL INSTRUMENTS.* Including—I. Instruments Used for Determining the Geometrical Features of a portion of Ground. II. Instruments Employed in Astronomical Observations. By J. F. HEATHER, M.A. Illustrated. 1s. 6d.

*** *The above three volumes form an enlargement of the Author's original work,* "*Mathematical Instruments.*" (*See No.* 32 *in the Series.*).

168. ⎫
169. ⎬ *MATHEMATICAL INSTRUMENTS.* By J. F. HEATHER,
170. ⎭ M.A. Enlarged Edition, for the most part entirely re-written. The 3 Parts as above, in One thick Volume. With numerous Illustrations. 4s. 6d.‡

158. *THE SLIDE RULE, AND HOW TO USE IT;* containing full, easy, and simple Instructions to perform all Business Calculations with unexampled rapidity and accuracy. By CHARLES HOARE, C.E. Sixth Edition. With a Slide Rule in tuck of cover. 2s. 6d.‡

196. *THEORY OF COMPOUND INTEREST AND ANNUI-TIES*; with Tables of Logarithms for the more Difficult Computations of Interest, Discount, Annuities, &c. By FÉDOR THOMAN. Fourth Edition. 4s.‡

199. *THE COMPENDIOUS CALCULATOR ;* or, Easy and Concise Methods of Performing the various Arithmetical Operations required in Commercial and Business Transactions; together with Useful Tables. By D. O'GORMAN. Twenty-seventh Edition, carefully revised by C. NORRIS. 2s. 6d., cloth limp; 3s. 6d., strongly half-bound in leather.

204. *MATHEMATICAL TABLES,* for Trigonometrical, Astronomical, and Nautical Calculations; to which is prefixed a Treatise on Logarithms. By HENRY LAW, C.E. Together with a :Series of Tables for Navigation and Nautical Astronomy. By Prof. J. R. YOUNG. New Edition. 4s.

204*. *LOGARITHMS.* With Mathematical Tables for Trigonometrical, Astronomical, and Nautical Calculations. By HENRY LAW, M.Inst.C.E. New and Revised Edition. (Forming part of the above Work). 3s.

221. *MEASURES, WEIGHTS, AND MONEYS OF ALL NA-TIONS,* and an Analysis of the Christian, Hebrew, and Mahometan Calendars. By W. S. B. WOOLHOUSE, F.R.A.S., F.S.S. Seventh Edition, 2s. 6d.‡

227. *MATHEMATICS AS APPLIED TO THE CONSTRUC-TIVE ARTS.* Illustrating the various processes of Mathematical Investigation, by means of Arithmetical and Simple Algebraical Equations and Practical Examples. By FRANCIS CAMPIN, C.E. Third Edition. 3s.‡

PHYSICAL SCIENCE, NATURAL PHILO-SOPHY, ETC.

1. *CHEMISTRY.* By Professor GEORGE FOWNES, F.R.S. With an Appendix on the Application of Chemistry to Agriculture. 1s.

2. *NATURAL PHILOSOPHY,* Introduction to the Study of. By C. TOMLINSON. Woodcuts. 1s. 6d.

6. *MECHANICS,* Rudimentary Treatise on. By CHARLES TOM-LINSON. Illustrated. 1s. 6d.

7. *ELECTRICITY;* showing the General Principles of Electrical Science, and the purposes to which it has been applied. By Sir W. SNOW HARRIS, F.R.S., &c. With Additions by R. SABINE, C.E., F.S.A. 1s. 6d.

7*. *GALVANISM.* By Sir W. SNOW HARRIS. New Edition by ROBERT SABINE, C.E., F.S.A. 1s. 6d.

8. *MAGNETISM;* being a concise Exposition of the General Principles of Magnetical Science. By Sir W. SNOW HARRIS. New Edition, revised by H. M. NOAD, Ph.D. With 165 Woodcuts. 3s. 6d.‡

☞ *The* ‡ *indicates that these vols. may be had strongly bound at 6d. extra.*

LONDON : CROSBY LOCKWOOD AND SON,

Physical Science, Natural Philosophy, etc., *continued.*

11. *THE ELECTRIC TELEGRAPH;* its History and Progress; with Descriptions of some of the Apparatus. By R. Sabine, C.E., F.S.A. 3s.

12. *PNEUMATICS,* including Acoustics and the Phenomena of Wind Currents, for the Use of Beginners By Charles Tomlinson, F.R.S. Fourth Edition, enlarged. Illustrated. 1s. 6d.

72. *MANUAL OF THE MOLLUSCA;* a Treatise on Recent and Fossil Shells. By Dr. S. P. Woodward, A.L.S. Fourth Edition. With Plates and 300 Woodcuts. 7s. 6d., cloth.

96. *ASTRONOMY.* By the late Rev. Robert Main, M.A. Third Edition, by William Thynne Lynn, B.A., F.R.A.S. 2s.

97. *STATICS AND DYNAMICS,* the Principles and Practice of; embracing also a clear development of Hydrostatics, Hydrodynamics, and Central Forces. By T. Baker, C.E. Fourth Edition. 1s. 6d.

173. *PHYSICAL GEOLOGY,* partly based on Major-General Portlock's "Rudiments of Geology." By Ralph Tate, A.L.S., &c. Woodcuts. 2s.

174. *HISTORICAL GEOLOGY,* partly based on Major-General Portlock's "Rudiments." By Ralph Tate, A.L.S., &c. Woodcuts. 2s. 6d.

173 *RUDIMENTARY TREATISE ON GEOLOGY,* Physical and
& Historical. Partly based on Major-General Portlock's "Rudiments of
174. Geology." By Ralph Tate, A.L.S., F.G.S., &c. In One Volume. 4s. 6d.‡

183 *ANIMAL PHYSICS,* Handbook of. By Dr. Lardner, D.C.L.,
& formerly Professor of Natural Philosophy and Astronomy in University
184. College, Lond. With 520 Illustrations. In One Vol. 7s. 6d., cloth boards.
*** *Sold also in Two Parts, as follows :—*
183. Animal Physics. By Dr. Lardner. Part I., Chapters I.—VII. 4s.
184. Animal Physics. By Dr. Lardner. Part II., Chapters VIII.—XVIII. 3s.

269. *LIGHT :* an Introduction to the Science of Optics, for the Use of Students of Architecture, Engineering, and other Applied Sciences. By E. Wyndham Tarn, M.A. 1s. 6d.

FINE ARTS.

20. *PERSPECTIVE FOR BEGINNERS.* Adapted to Young Students and Amateurs in Architecture, Painting, &c. By George Pyne. 2s.

40 *GLASS STAINING, AND THE ART OF PAINTING ON GLASS.* From the German of Dr. Gessert and Emanuel Otto Fromberg. With an Appendix on The Art of Enamelling. 2s 6d.

69. *MUSIC,* A Rudimentary and Practical Treatise on. With numerous Examples. By Charles Child Spencer. 2s. 6d.

71. *PIANOFORTE,* The Art of Playing the. With numerous Exercises & Lessons from the Best Masters. By Charles Child Spencer. 1s.6d.

69-71. *MUSIC & THE PIANOFORTE.* In one vol. Half bound, 5s.

181. *PAINTING POPULARLY EXPLAINED,* including Fresco, Oil, Mosaic, Water Colour, Water-Glass, Tempera, Encaustic, Miniature, Painting on Ivory, Vellum, Pottery, Enamel, Glass, &c. With Historical Sketches of the Progress of the Art by Thomas John Gullick, assisted by John Timbs, F.S.A. Sixth Edition, revised and enlarged. 5s.‡

186. *A GRAMMAR OF COLOURING,* applied to Decorative Painting and the Arts. By George Field. New Edition, enlarged and adapted to the Use of the Ornamental Painter and Designer. By Ellis A. Davidson. With two new Coloured Diagrams, &c. 3s.‡

246. *A DICTIONARY OF PAINTERS, AND HANDBOOK FOR PICTURE AMATEURS;* including Methods of Painting, Cleaning, Relining and Restoring, Schools of Painting, &c. With Notes on the Copyists and Imitators of each Master. By Philippe Daryl. 2s. 6d.‡

☞ *The ‡ indicates that these vols. may be had strongly bound at 6d. extra.*

INDUSTRIAL AND USEFUL ARTS.

23. *BRICKS AND TILES*, Rudimentary Treatise on the Manufacture of. By E. DOBSON, M.R.I.B.A. Illustrated, 3s.‡

67. *CLOCKS, WATCHES, AND BELLS*, a Rudimentary Treatise on. By Sir EDMUND BECKETT, LL.D., Q.C. Seventh Edition, revised and enlarged. 4s. 6d. limp; 5s. 6d. cloth boards.

83**. *CONSTRUCTION OF DOOR LOCKS.* Compiled from the Papers of A. C. HOBBS, and Edited by CHARLES TOMLINSON. F.R.S. 2s. 6d.

162. *THE BRASS FOUNDER'S MANUAL;* Instructions for Modelling, Pattern-Making, Moulding, Turning, Filing, Burnishing, Bronzing, &c. With copious Receipts, &c. By WALTER GRAHAM. 2s.‡

205. *THE ART OF LETTER PAINTING MADE EASY.* By J. G. BADENOCH. Illustrated with 12 full-page Engravings of Examples. 1s. 6d.

215. *THE GOLDSMITH'S HANDBOOK*, containing full Instructions for the Alloying and Working of Gold. By GEORGE E. GEE, 3s.‡

225. *THE SILVERSMITH'S HANDBOOK*, containing full Instructions for the Alloying and Working of Silver. By GEORGE E. GEE. 3s.‡
⁎ The two preceding Works, in One handsome Vol., half-bound, entitled "THE GOLDSMITH'S & SILVERSMITH'S COMPLETE HANDBOOK," 7s.

249. *THE HALL-MARKING OF JEWELLERY PRACTICALLY CONSIDERED.* By GEORGE E. GEE. 3s.‡

224. *COACH BUILDING*, A Practical Treatise, Historical and Descriptive. By J. W. BURGESS. 2s. 6d.‡

235. *PRACTICAL ORGAN BUILDING.* By W. E. DICKSON, M.A., Precentor of Ely Cathedral. Illustrated. 2s. 6d.‡

262. *THE ART OF BOOT AND SHOEMAKING.* By JOHN BEDFORD LENO. Numerous Illustrations. Third Edition. 2s.

263. *MECHANICAL DENTISTRY:* A Practical Treatise on the Construction of the Various Kinds of Artificial Dentures, with Formulæ, Tables, Receipts, &c. By CHARLES HUNTER. Third Edition. 3s.‡

270. *WOOD ENGRAVING:* A Practical and Easy Introduction to the Study of the Art. By W. N. BROWN. 1s. 6d.

MISCELLANEOUS VOLUMES.

36. *A DICTIONARY OF TERMS used in ARCHITECTURE, BUILDING, ENGINEERING, MINING, METALLURGY, ARCHÆOLOGY, the FINE ARTS, &c.* By JOHN WEALE. Sixth Edition. Revised by ROBERT HUNT, F.R.S. Illustrated. 5s. limp; 6s. cloth boards.

50. *LABOUR CONTRACTS.* A Popular Handbook on the Law of Contracts for Works and Services. By DAVID GIBBONS. Fourth Edition, Revised, with Appendix of Statutes by T. F. UTILEY, Solicitor, 3s. 6d. cloth.

112. *MANUAL OF DOMESTIC MEDICINE.* By R. GOODING, B.A., M.D. A Family Guide in all Cases of Accident and Emergency 2s.

112*. *MANAGEMENT OF HEALTH.* A Manual of Home and Personal Hygiene. By the Rev. JAMES BAIRD, B.A. 1s.

150. *LOGIC*, Pure and Applied. By S. H. EMMENS. 1s. 6d.

153. *SELECTIONS FROM LOCKE'S ESSAYS ON THE HUMAN UNDERSTANDING.* With Notes by S. H. EMMENS. 2s.

154. *GENERAL HINTS TO EMIGRANTS.* 2s.

157. *THE EMIGRANT'S GUIDE TO NATAL.* By R. MANN. 2s.

193. *HANDBOOK OF FIELD FORTIFICATION.* By Major W. W. KNOLLYS, F.R.G.S. With 163 Woodcuts. 3s.‡

194. *THE HOUSE MANAGER:* Being a Guide to Housekeeping. Practical Cookery, Pickling and Preserving, Household Work, Dairy Management, &c. By AN OLD HOUSEKEEPER. 3s. 6d.‡

194, *HOUSE BOOK* (*The*). Comprising :—I. THE HOUSE MANAGER.
112 & By an OLD HOUSEKEEPER. II. DOMESTIC MEDICINE. By R. GOODING, M.D.
112*. III. MANAGEMENT OF HEALTH. By J. BAIRD. In One Vol., half-bound, 6s.

☞ *The ‡ indicates that these vols. may be had strongly bound at 6d. extra.*

LONDON : CROSBY LOCKWOOD AND SON,

EDUCATIONAL AND CLASSICAL SERIES.

HISTORY.

1. **England, Outlines of the History of;** more especially with reference to the Origin and Progress of the English Constitution. By WILLIAM DOUGLAS HAMILTON, F.S.A., of Her Majesty's Public Record Office. 4th Edition, revised. 5s.; cloth boards, 6s.

5. **Greece, Outlines of the History of;** in connection with the Rise of the Arts and Civilization in Europe. By W. DOUGLAS HAMILTON, of University College, London, and EDWARD LEVIEN, M.A., of Balliol College, Oxford. 2s. 6d.; cloth boards, 3s. 6d.

7. **Rome, Outlines of the History of:** from the Earliest Period to the Christian Era and the Commencement of the Decline of the Empire. By EDWARD LEVIEN, of Balliol College, Oxford. Map, 2s. 6d.; cl. bds. 3s. 6d.

9. **Chronology of History, Art, Literature, and Progress,** from the Creation of the World to the Present Time. The Continuation by W. D. HAMILTON, F.S.A. 3s.; cloth boards, 3s. 6d.

ENGLISH LANGUAGE AND MISCELLANEOUS.

11. **Grammar of the English Tongue,** Spoken and Written. With an Introduction to the Study of Comparative Philology. By HYDE CLARKE, D.C.L. Fifth Edition. 1s. 6d.

12. **Dictionary of the English Language,** as Spoken and Written. Containing above 100,000 Words. By HYDE CLARKE, D.C.L. 3s. 6d.; cloth boards, 4s. 6d.; complete with the GRAMMAR, cloth bds., 5s. 6d.

48. **Composition and Punctuation,** familiarly Explained for those who have neglected the Study of Grammar. By JUSTIN BRENAN. 18th Edition. 1s. 6d.

49. **Derivative Spelling-Book:** Giving the Origin of Every Word from the Greek, Latin, Saxon, German, Teutonic, Dutch, French, Spanish, and other Languages; with their present Acceptation and Pronunciation. By J. ROWBOTHAM, F.R.A.S. Improved Edition. 1s. 6d.

The Art of Extempore Speaking: Hints for the Pulpit, the Senate, and the Bar. By M. BAUTAIN, Vicar-General and Professor at the Sorbonne. Translated from the French. 8th Edition, carefully corrected. 2s. 6d.

53. **Places and Facts in Political and Physical Geography,** for Candidates in Examinations. By the Rev. EDGAR RAND, B.A. 1s.

54. **Analytical Chemistry,** Qualitative and Quantitative, a Course of. To which is prefixed, a Brief Treatise upon Modern Chemical Nomenclature and Notation. By WM. W. PINK and GEORGE E. WEBSTER. 2s.

THE SCHOOL MANAGERS' SERIES OF READING BOOKS,

Edited by the Rev. A. R. GRANT, Rector of Hitcham, and Honorary Canon of Ely; formerly H.M. Inspector of Schools. INTRODUCTORY PRIMER, 3d.

	s.	d.				s.	d.
FIRST STANDARD .	. 0	6	FOURTH STANDARD	.	. .	1	2
SECOND ,,	. . 0	10	FIFTH ,,	.	. .	1	6
THIRD ,,	. . 1	0	SIXTH ,,	.	. .	1	6

LESSONS FROM THE BIBLE. Part I. Old Testament. 1s.

LESSONS FROM THE BIBLE. Part II. New Testament, to which is added THE GEOGRAPHY OF THE BIBLE, for very young Children. By Rev. C. THORNTON FORSTER. 1s. 2d. *.* Or the Two Parts in One Volume. 2s.

7, STATIONERS' HALL COURT, LUDGATE HILL, E C.

FRENCH.

24. French Grammar. With Complete and Concise Rules on the Genders of French Nouns. By G. L. STRAUSS, Ph.D. 1s. 6d.

25. French-English Dictionary. Comprising a large number of New Terms used in Engineering, Mining, &c. By ALFRED ELWES. 1s. 6d.

26. English-French Dictionary. By ALFRED ELWES. 2s.

25,26. French Dictionary (as above). Complete, in One Vol., 3s.; cloth boards, 3s. 6d. *₊* Or with the GRAMMAR, cloth boards, 4s. 6d.

47. French and English Phrase Book : containing Introductory Lessons, with Translations, several Vocabularies of Words a Collection of suitable Phrases, and Easy Familiar Dialogues. 1s. 6d.

GERMAN.

39. German Grammar. Adapted for English Students, from Heyse's Theoretical and Practical Grammar, by Dr. G. L. STRAUSS. 1s. 6d.

40. German Reader : A Series of Extracts, carefully culled from the most approved Authors of Germany; with Notes, Philological and Explanatory. By G. L. STRAUSS, Ph.D. 1s.

41-43. German Triglot Dictionary. By N. E. S. A. HAMILTON. In Three Parts. Part I. German-French-English. Part II. English-German-French. Part III. French-German-English. 3s., or cloth boards, 4s.

41-43 & 39. German Triglot Dictionary (as above), together with German Grammar (No. 39). in One Volume, cloth boards, 5s.

ITALIAN.

27. Italian Grammar, arranged in Twenty Lessons, with a Course of Exercises. By ALFRED ELWES. 1s. 6d.

28. Italian Triglot Dictionary, wherein the Genders of all the Italian and French Nouns are carefully noted down. By ALFRED ELWES. Vol. 1. Italian-English-French. 2s. 6d.

30. Italian Triglot Dictionary. By A. ELWES. Vol. 2. English-French-Italian. 2s. 6d.

32. Italian Triglot Dictionary. By ALFRED ELWES. Vol. 3. French-Italian-English. 2s. 6d.

28,30, 32. Italian Triglot Dictionary (as above). In One Vol., 7s. 6d. Clo h boards.

SPANISH AND PORTUGUESE.

34. Spanish Grammar, in a Simple and Practical Form. With a Course of Exercises. By ALFRED ELWES. 1s. 6d.

35. Spanish-English and English-Spanish Dictionary. Including a large number of Technical Terms used in Mining, Engineering, &c. with the proper Accents and the Gender of every Noun. By ALFRED ELWES 4s.; cloth boards, 5s. *₊* Or with the GRAMMAR, cloth boards, 6s.

55. Portuguese Grammar, in a Simple and Practical Form. With a Course of Exercises. By ALFRED ELWES. 1s. 6d.

56. Portuguese-English and English-Portuguese Dictionary. Including a large number of Technical Terms used in Mining, Engineering, &c., with the proper Accents and the Gender of every Noun. By ALFRED ELWES. Third Edition, Revised, 5s.; cloth boards, 6s. *₊* Or with the GRAMMAR, cloth boards, 7s.

HEBREW.

46*. Hebrew Grammar. By Dr. BRESSLAU. 1s. 6d.

44. Hebrew and English Dictionary, Biblical and Rabbinical; containing the Hebrew and Chaldee Roots of the Old Testament Post-Rabbinical Writings. By Dr. BRESSLAU. 6s.

46. English and Hebrew Dictionary. By Dr. BRESSLAU. 3s.

44,46. 46*. Hebrew Dictionary (as above), in Two Vols., complete, with the GRAMMAR, cloth boards, 12s.

LONDON : CROSBY LOCKWOOD AND SON,

LATIN.

19. **Latin Grammar.** Containing the Inflections and Elementary Principles of Translation and Construction. By the Rev. THOMAS GOODWIN, M.A., Head Master of the Greenwich Proprietary School. 1s. 6d.

20. **Latin-English Dictionary.** By the Rev. THOMAS GOODWIN, M.A. 2s.

22. **English-Latin Dictionary;** together with an Appendix of French and Italian Words which have their origin from the Latin. By the Rev. THOMAS GOODWIN, M.A. 1s. 6d.

20,22. **Latin Dictionary** (as above). Complete in One Vol., 3s. 6d. cloth boards, 4s. 6d. ** Or with the GRAMMAR, cloth boards, 5s. 6d.

LATIN CLASSICS. With Explanatory Notes in English.

1. **Latin Delectus.** Containing Extracts from Classical Authors, with Genealogical Vocabularies and Explanatory Notes, by H. YOUNG. 1s. 6d:

2. **Cæsaris** Commentarii de Bello Gallico. Notes, and a Geographical Register for the Use of Schools, by H. YOUNG. 2s.

3. **Cornelius Nepos.** With Notes. By H. YOUNG. 1s.

4. **Virgilii** Maronis Bucolica et Georgica. With Notes on the Bucolics by W. RUSHTON, M.A., and on the Georgics by H. YOUNG. 1s. 6d.

5. **Virgilii Maronis Æneis.** With Notes, Critical and Explanatory, by H. YOUNG. New Edition, revised and improved. With copious Additional Notes by Rev. T. H. L. LEARY, D.C.L., formerly Scholar of Brasenose College, Oxford. 3s.

5*. ———— Part 1. Books i.—vi., 1s. 6d.
5**. ———— Part 2. Books vii.—xii., 2s.

6. **Horace;** Odes, Epode, and Carmen Sæculare. Notes by H. YOUNG. 1s. 6d.

7. **Horace;** Satires, Epistles, and Ars Poetica. Notes by W. BROWNRIGG SMITH, M.A., F.R.G.S. 1s. 6d.

8. **Sallustii** Crispi Catalina et Bellum Jugurthinum. Notes, Critical and Explanatory, by W. M. DONNE, B.A., Trin. Coll., Cam. 1s. 6d.

9. **Terentii** Andria et Heautontimorumenos. With Notes, Critical and Explanatory, by the Rev. JAMES DAVIES, M.A. 1s. 6d.

10. **Terentii** Adelphi, Hecyra, Phormio. Edited, with Notes, Critical and Explanatory, by the Rev. JAMES DAVIES, M.A. 2s.

11. **Terentii** Eunuchus, Comœdia. Notes, by Rev. J. DAVIES, M.A. 1s. 6d.

12. **Ciceronis** Oratio pro Sexto Roscio Amerino. Edited, with an Introduction, Analysis, and Notes, Explanatory and Critical, by the Rev. JAMES DAVIES, M.A. 1s. 6d.

13. **Ciceronis** Orationes in Catilinam, Verrem, et pro Archia. With Introduction. Analysis, and Notes, Explanatory and Critical, by Rev. T. H. L. LEARY, D.C.L. formerly Scholar of Brasenose College, Oxford. 1s. 6d.

14. **Ciceronis** Cato Major, Lælius, Brutus, sive de Senectute, de Amicitia, de Claris Oratoribus Dialogi. With Notes by W. BROWNRIGG SMITH M.A., F.R.G.S. 2s.

16. **Livy:** History of Rome. Notes by H. YOUNG and W. B. SMITH, M.A. Part 1. Books i., ii., 1s. 6d.

16*. ———— Part 2. Books iii., iv., v., 1s. 6d.
17. ———— Part 3. Books xxi., xxii., 1s. 6d.

19. **Latin Verse Selections,** from Catullus, Tibullus, Propertius, and Ovid. Notes by W. B. DONNE, M.A., Trinity College, Cambridge. 2s.

20. **Latin Prose Selections,** from Varro, Columella, Vitruvius, Seneca, Quintilian, Florus, Velleius Paterculus, Valerius Maximus Suetonius, Apuleius, &c. Notes by W. B. DONNE, M.A. 2s.

21. **Juvenalis** Satiræ. With Prolegomena and Notes by T. H. S. ESCOTT. B.A., Lecturer on Logic at King's College, London. 2s.

GREEK.

14. Greek Grammar, in accordance with the Principles and Philological Researches of the most eminent Scholars of our own day. By HANS CLAUDE HAMILTON. 1s. 6d.

15,17. Greek Lexicon. Containing all the Words in General Use, with their Significations, Inflections, and Doubtful Quantities. By HENRY R. HAMILTON. Vol. 1. Greel -English, 2s. 6d.; Vol. 2. English-Greek, 2s. Or the Two Vols. in One, 4s. 6d.: cloth boards, 5s.

14,15. Greek Lexicon (as above). Complete, with the GRAMMAR, in **17.** One Vol., cloth boards, 6s.

GREEK CLASSICS.   With Explanatory Notes in English.

1. Greek Delectus. Containing Extracts from Classical Authors, with Genealogical Vocabularies and Explanatory Notes, by H. YOUNG. New Edition, with an improved and enlarged Supplementary Vocabulary, by JOHN HUTCHISON, M.A., of the High School, Glasgow. 1s. 6d.

2, 3. Xenophon's Anabasis; or, The Retreat of the Ten Thousand. Notes and a Geographical Register, by H. YOUNG. Part 1. Books i. to iii., 1s. Part 2. Books iv. to vii., 1s.

4. Lucian's Select Dialogues. The Text carefully revised, with Grammatical and Explanatory Notes, by H. YOUNG. 1s. 6d.

5-12. Homer, The Works of. According to the Text of BAEUMLEIN. With Notes, Critical and Explanatory, drawn from the best and latest Authorities, with Preliminary Observations and Appendices, by T. H. L. LEARY, M.A., D.C.L.

THE ILIAD:	Part 1. Books i. to vi., 1s.6d.	Part 3. Books xiii. to xviii., 1s. 6d
	Part 2. Books vii. to xii., 1s.6d.	Part 4. Books xix. to xxiv., 1s. 6d.
THE ODYSSEY:	Part 1. Books i. to vi., 1s. 6d	Part 3. Books xiii. to xviii., 1s. 6d.
	Part 2. Books vii. to xii., 1s.6d.	Part 4. Books xix. to xxiv., and Hymns, 2s.

13. Plato's Dialogues: The Apology of Socrates, the Crito, and the Phædo. From the Text of C. F. HERMANN. Edited with Notes, Critical and Explanatory, by the Rev. JAMES DAVIES, M.A. 2s.

14-17. Herodotus, The History of, chiefly after the Text of GAISFORD. With Preliminary Observations and Appendices, and Notes, Critical and Explanatory, by T. H. L. LEARY, M.A., D.C.L.
Part 1.   Books i., ii. (The Clio and Euterpe), 2s.
Part 2.   Books iii., iv. (The Thalia and Melpomene), 2s.
Part 3.   Books v.-vii. (The Terpsichore, Erato, and Polymnia), 2s.
Part 4.   Books viii., ix. (The Urania and Calliope) and Index, 1s. 6d.

18. Sophocles: Œdipus Tyrannus. Notes by H. YOUNG. 1s.

20. Sophocles: Antigone. From the Text of DINDORF. Notes, Critical and Explanatory, by the Rev. JOHN MILNER, B.A. 2s.

23. Euripides: Hecuba and Medea. Chiefly from the Text of DINDORF. With Notes, Critical and Explanatory, by W. BROWNRIGG SMITH, M.A., F.R.G.S. 1s. 6d.

26. Euripides: Alcestis. Chiefly from the Text of DINDORF. With Notes, Critical and Explanatory, by JOHN MILNER, B.A. 1s. 6d.

30. Æschylus: Prometheus Vinctus: The Prometheus Bound. From the Text of DINDORF. Edited, with English Notes, Critical and Explanatory, by the Rev. JAMES DAVIES, M.A. 1s.

32. Æschylus: Septem 'Contra Thebes: The Seven against Thebes. From the Text of DINDORF. Edited, with English Notes, Critical and Explanatory, by the Rev. JAMES DAVIES, M.A. 1s.

40. Aristophanes: Acharnians. Chiefly from the Text of C. H WEISE. With Notes, by C. S. T. TOWNSHEND, M.A. 1s. 6d.

41. Thucydides: History of the Peloponnesian War. Notes by H. YOUNG. Book 1. 1s. 6d.

42. Xenophon's Panegyric on Agesilaus. Notes and Introduction by LL. F. W. JEWITT. 1s. 6d.

43. Demosthenes. The Oration on the Crown and the Philippics. With English Notes. By Rev. T. H. L. LEARY, D.C.L., formerly Scholar of Brasenose College, Oxford. 1s. 6d.

7, STATIONERS' HALL COURT, LONDON, E.C.

September, 1896.

A
CATALOGUE OF BOOKS
INCLUDING NEW AND STANDARD WORKS IN
ENGINEERING: CIVIL, MECHANICAL, AND MARINE;
ELECTRICITY AND ELECTRICAL ENGINEERING;
MINING, METALLURGY; ARCHITECTURE,
BUILDING, INDUSTRIAL AND DECORATIVE ARTS;
SCIENCE, TRADE AND MANUFACTURES;
AGRICULTURE, FARMING, GARDENING;
AUCTIONEERING, VALUING AND ESTATE AGENCY;
LAW AND MISCELLANEOUS.

PUBLISHED BY
CROSBY LOCKWOOD & SON.

MECHANICAL ENGINEERING, etc.

D. K. Clark's Pocket-Book for Mechanical Engineers.
THE MECHANICAL ENGINEER'S POCKET-BOOK OF *TABLES, FORMULÆ, RULES AND DATA.* A Handy Book of Reference for Daily Use in Engineering Practice. By D. KINNEAR CLARK, M.Inst.C.E., Author of "Railway Machinery," "Tramways," &c. Third Edition, Revised. Small 8vo, 700 pages, 6s. bound in flexible leather cover, rounded corners.

SUMMARY OF CONTENTS.
MATHEMATICAL TABLES.—MEASUREMENT OF SURFACES AND SOLIDS.— ENGLISH WEIGHTS AND MEASURES.—FRENCH METRIC WEIGHTS AND MEASURES.—FOREIGN WEIGHTS AND MEASURES.—MONEYS.—SPECIFIC GRAVITY, WEIGHT AND VOLUME.—MANUFACTURED METALS.—STEEL PIPES.—BOLTS AND NUTS.—SUNDRY ARTICLES IN WROUGHT AND CAST IRON, COPPER, BRASS, LEAD, TIN, ZINC.—STRENGTH OF MATERIALS.—STRENGTH OF TIMBER.—STRENGTH OF CAST IRON.—STRENGTH OF WROUGHT IRON.—STRENGTH OF STEEL.—TENSILE STRENGTH OF COPPER, LEAD, ETC.—RESISTANCE OF STONES AND OTHER BUILDING MATERIALS.—RIVETED JOINTS IN BOILER PLATES.—BOILER SHELLS.—WIRE ROPES AND HEMP ROPES.—CHAINS AND CHAIN CABLES.—FRAMING.—HARDNESS OF METALS, ALLOYS AND STONES.—LABOUR OF ANIMALS.—MECHANICAL PRINCIPLES.—GRAVITY AND FALL OF BODIES.—ACCELERATING AND RETARDING FORCES.—MILL GEARING, SHAFTING, ETC.—TRANSMISSION OF MOTIVE POWER.— HEAT.—COMBUSTION: FUELS.—WARMING, VENTILATION, COOKING STOVES.— STEAM.—STEAM ENGINES AND BOILERS.—RAILWAYS.—TRAMWAYS.—STEAM SHIPS.—PUMPING STEAM ENGINES AND PUMPS.—COAL GAS, GAS ENGINES, ETC.— AIR IN MOTION.—COMPRESSED AIR.—HOT AIR ENGINES.—WATER POWER.— SPEED OF CUTTING TOOLS.—COLOURS.—ELECTRICAL ENGINEERING.

*** OPINIONS OF THE PRESS.

" Mr. Clark manifests what is an innate perception of what is likely to be useful in a pocket-book, and he is really unrivalled in the art of condensation. Very frequently we find the information on a given subject is supplied by giving a summary description of an experiment, and a statement of the results obtained. There is a very excellent steam table, occupying five and-a-half pages ; and there are rules given for several calculations, which rules cannot be found in other pocket-books, as, for example, that on page 497, for getting at the quantity of water in the shape of priming in any known weight of steam. It is very difficult to hit upon any mechanical engineering subject concerning which this work supplies no information, and the excellent index at the end adds to its utility. In one word, it is an exceedingly handy and efficient tool, possessed of which the engineer will be saved many a wearisome calculation, or yet more wearisome hunt through various text-books and treatises, and, as such, we can heartily recommend it to our readers, who must not run away with the idea that 'Mr. Clark's Pocket-book is only Molesworth in another form. On the contrary, each contains what is not to be found in the other ; and Mr. Clark takes more room and deals at more length with many subjects than Molesworth possibly could."
The Engineer.

" It would be found difficult to compress more matter within a similar compass, or produce a book of 650 pages which should be more compact or convenient for pocket reference. . . . Will be appreciated by mechnaical engineers of all classes."—*Practical Engineer.*

" Just the kind of work that practical men require to have near to them."—*English Mechanic,*

B

MR. HUTTON'S PRACTICAL HANDBOOKS.

Handbook for Works' Managers.

THE WORKS' MANAGER'S HANDBOOK OF MODERN RULES, TABLES, AND DATA. For Engineers, Millwrights, and Boiler Makers; Tool Makers, Machinists, and Metal Workers; Iron and Brass Founders, &c. By W. S. HUTTON, Civil and Mechanical Engineer, Author of "The Practical Engineer's Handbook." Fifth Edition, carefully Revised, with Additions. In One handsome Volume, medium 8vo, price 15*s*. strongly bound.   [*Just published.*

☞ *The Author having compiled Rules and Data for his own use in a great variety of modern engineering work, and having found his notes extremely useful, decided to publish them—revised to date—believing that a practical work, suited to the* DAILY REQUIREMENTS OF MODERN ENGINEERS, *would be favourably received.*

In the Fourth Edition the First Section has been re-written and improved by the addition of numerous Illustrations and new matter relating to STEAM ENGINES *and* GAS ENGINES. *The Second Section has been enlarged and Illustrated, and throughout the book a great number of emendations and alterations have been made, with the object of rendering the book more generally useful.*

*** OPINIONS OF THE PRESS.

"The author treats every subject from the point of view of one who has collected workshop notes for application in workshop practice, rather than from the theoretical or literary aspect. The volume contains a great deal of that kind of information which is gained only by practical experience, and is seldom written in books."—*Engineer.*

"The volume is an exceedingly useful one, brimful with engineers' notes, memoranda, and rules, and well worthy of being on every mechanical engineer's bookshelf."—*Mechanical World.*

"The information is precisely that likely to be required in practice. . . . The work forms a desirable addition to the library not only of the works' manager, but of anyone connected with general engineering."—*Mining Journal.*

"A formidable mass of facts and figures, readily accessible through an elaborate index . . . Such a volume will be found absolutely necessary as a book of reference in all sorts of 'works' connected with the metal trades."—*Ryland's Iron Trades Circular.*

"Brimful of useful information, stated in a concise form, Mr. Hutton's books have met a pressing want among engineers. The book must prove extremely useful to every practical man possessing a copy."—*Practical Engineer.*

New Manual for Practical Engineers.

THE PRACTICAL ENGINEER'S HAND-BOOK. Comprising a Treatise on Modern Engines and Boilers: Marine, Locomotive and Stationary. And containing a large collection of Rules and Practical Data relating to recent Practice in Designing and Constructing all kinds of Engines, Boilers, and other Engineering work. The whole constituting a comprehensive Key to the Board of Trade and other Examinations for Certificates of Competency in Modern Mechanical Engineering. By WALTER S. HUTTON, Civil and Mechanical Engineer, Author of "The Works' Manager's Handbook for Engineers," &c. With upwards of 370 Illustrations. Fifth Edition, Revised, with Additions. Medium 8vo, nearly 500 pp., price 18*s*. Strongly bound.   [*Just published.*

☞ *This work is designed as a companion to the Author's* "WORKS' MANAGER'S HAND-BOOK." *It possesses many new and original features, and contains, like its predecessor, a quantity of matter not originally intended for publication, but collected by the author for his own use in the construction of a great variety of* MODERN ENGINEERING WORK.

The information is given in a condensed and concise form, and is illustrated by upwards of 370 Woodcuts; and comprises a quantity of tabulated matter of great value to all engaged in designing, constructing, or estimating for ENGINES, BOILERS, *and* OTHER ENGINEERING WORK.

*** OPINIONS OF THE PRESS.

"We have kept it at hand for several weeks, referring to it as occasion arose, and we have not on a single occasion consulted its pages without finding the information of which we were in quest." —*Athenæum.*

"A thoroughly good practical handbook, which no engineer can go through without learning something that will be of service to him."—*Marine Engineer.*

"An excellent book of reference for engineers, and a valuable text-book for students of engineering."—*Scotsman.*

"This valuable manual embodies the results and experience of the leading authorities on mechanical engineering."—*Building News.*

"The author has collected together a surprising quantity of rules and practical data, and has shown much judgment in the selections he has made. . . . There is no doubt that this book is one of the most useful of its kind published, and will be a very popular compendium."—*Engineer.*

"A mass of information, set down in simple language, and in such a form that it can be easily referred to at any time. The matter is uniformly good and well chosen and is greatly elucidated by the illustrations. The book will find its way on to most engineers' shelves, where it will rank as one of the most useful books of reference."—*Practical Engineer.*

"Full of useful information and should be found on the office shelf of all practical engineers." —*English Mechanic.*

MR. HUTTON'S PRACTICAL HANDBOOKS—*continued.*

Practical Treatise on Modern Steam-Boilers.

STEAM-BOILER CONSTRUCTION. A Practical Handbook for Engineers, Boiler-Makers, and Steam Users. Containing a large Collection of Ru!es and Data relating to Recent Practice in the Design, Construction, and Working of all Kinds of Stationary, Locomotive, and Marine Steam-Boilers. By WALTER S. HUTTON, Civil and Mechanical Engineer, Author of "The Works' Manager's Handbook," "The Practical Engineer's Handbook," &c. With upwards of 300 Illustrations. Second Edition. Medium 8vo, 18s. cloth.

☞ *This work is issued in continuation of the Series of Handbooks written by the Author, viz:—"*THE WORKS' MANAGER'S HANDBOOK*" and "*THE PRACTICAL ENGINEER'S HANDBOOK,*" which are so highly appreciated by Engineers for the practical nature of their information; and is consequently written in the same style as those works.*

The Author believes that the concentration, in a convenient form for easy reference, of such a large amount of thoroughly practical information on Steam-Boilers, will be of considerable service to those for whom it is intended, and he trusts the book may be deemed worthy of as favourable a reception as has been accorded to its predecessors.

** OPINIONS OF THE PRESS.

"Every detail, both in boiler design and management, is clearly laid before the reader. The volume shows that boiler construction has been reduced to the condition of one of the most exact sciences; and such a book is of the utmost value to the *fin de siècle* Engineer and Works Manager."—*Marine Engineer.*

"There has long been room for a modern handbook on steam boilers; there is not that room now, because Mr. Hutton has filled it. It is a thoroughly practical book for those who are occupied in the construction, design, selection, or use of boilers."—*Engineer.*

"The book is of so important and comprehensive a character that it must find its way into the libraries of everyone interested in boiler using or boiler manufacture if they wish to be thoroughly informed. We strongly recommend the book for the intrinsic value of its contents."—*Machinery Market.*

"The value of this book can hardly be over-estimated The author's rules, formulæ &c., are all very fresh, and it is impossible to turn to the work and not find what you want. No practical engineer should be without it."—*Colliery Guardian.*

Hutton's "Modernised Templeton."

THE PRACTICAL MECHANICS' WORKSHOP COMPANION. Comprising a great variety of the most useful Ru!es and Formulæ in Mechanical Science, with numerous Tables of Practical Data and Calculated Results for Facilitating Mechanical Operations. By WILLIAM TEMPLETON, Author of "The Engineer's Practical Assistant," &c. &c. Seventeenth Edition, Revised, Modernised, and considerably Enlarged by WALTER S. HUTTON, C.E., Author of "The Works' Manager's Handbook," "The Practical Engineer's Handbook," &c. Fcap. 8vo, nearly 500 pp., with 8 Plates and upwards of 250 Illustrative Diagrams, 6s., strongly bound for workshop or pocket wear and tear.

** OPINIONS OF THE PRESS.

"In its modernised form Hutton's 'Templeton 'should have a wide sale, for it contains much valuable information which the mechanic will often find of use, and not a few tables and notes which he might look for in vain in other works. This modernised edition will be appreciated by all who have learned to value the original editions of ' Templeton.'' —*English Mechanic.*

"It has met with great success in the engineering workshop, as we can testify; and there are a great many men who, in a great measure, owe their rise in life to this little book."—*Building News.*

"This familiar text-book—well known to all mechanics and engineers—is of essential service to the every-day requirements of engineers, millwrights, and the various trades connected with engineering and building. The new modernised edition is worth its weight in gold."—*Building News.* (Second Notice.)

"This well-known and largely used book contains information, brought up to date, of the sort so useful to the foreman and draughtsman. So much fresh information has been introduced as to constitute it practically a new book. It will be largely used in the office and workshop."—*Mechanical World.*

"The publishers wisely entrusted the task of revision of this popular, valuable, and useful book to Mr. Hutton, than whom a more competent man they could not have found."—*Iron.*

Templeton's Engineer's and Machinist's Assistant.

THE ENGINEER'S, MILLWRIGHT'S, and MACHINIST'S PRACTICAL ASSISTANT. A collection of Useful Tables, Rules and Data. By WILLIAM TEMPLETON. 7th Edition, with Additions. 18mo, 2s. 6d. cloth.

** OPINIONS OF THE PRESS.

"Occupies a foremost place among books of this kind. A more suitable present to an apprentice to any of the mechanical trades could not possibly be made."—*Building News.*

"A deservedly popular work. It should be in the 'drawer' of every mechanic."—*English Mechanic.*

Foley's Office Reference Book for Mechanical Engineers.

THE MECHANICAL ENGINEER'S REFERENCE BOOK, for Machine and Boiler Construction. In Two Parts. Part I. GENERAL ENGINEERING DATA. Part II. BOILER CONSTRUCTION. With 51 Plates and numerous Illustrations. By NELSON FOLEY, M.I.N.A.   Second Edition, Revised throughout and much Enlarged. Folio, £3 3s. net half-bound.

SUMMARY OF CONTENTS.   [*Just published.*

PART I.

MEASURES.—CIRCUMFERENCES AND AREAS, &c., SQUARES, CUBES, FOURTH POWERS.—SQUARE AND CUBE ROOTS.— SURFACE OF TUBES—RECIPROCALS.— LOGARITHMS.— MENSURATION. — SPECIFIC GRAVITIES AND WEIGHTS.— WORK AND POWER.—HEAT.—COMBUSTION.—EXPANSION AND CONTRACTION. —EXPANSION OF GASES.—STEAM.— STATIC FORCES.—GRAVITATION AND ATTRACTION.—MOTION AND COMPUTATION OF RESULTING FORCES.—ACCUMULATED WORK.—CENTRE AND RADIUS OF GYRATION.—MOMENT OF INERTIA. —CENTRE OF OSCILLATION.—ELECTRICITY.—STRENGTH OF MATERIALS. —ELASTICITY. — TEST SHEETS OF METALS.— FRICTION. —TRANSMISSION OF POWER.—FLOW OF LIQUIDS.—FLOW OF GASES.—AIR PUMPS, SURFACE CONDENSERS, &c.—SPEED OF STEAMSHIPS. —PROPELLERS. — CUTTING TOOLS.— FLANGES. — COPPER SHEETS AND TUBES.—SCREWS, NUTS, BOLT HEADS, &c.—VARIOUS RECIPES AND MISCELLANEOUS MATTER.

WITH DIAGRAMS FOR VALVE-GEAR, BELTING AND ROPES, DISCHARGE AND SUCTION PIPES, SCREW PROPELLERS, AND COPPER PIPES.

PART II.

TREATING OF, POWER OF BOILERS.— USEFUL RATIOS.—NOTES ON CONSTRUCTION. — CYLINDRICAL BOILER SHELLS. — CIRCULAR FURNACES. — FLAT PLATES.— STAYS.— GIRDERS.— SCREWS. — HYDRAULIC TESTS. — RIVETING.—BOILER SETTING, CHIMNEYS, AND MOUNTINGS.—FUELS, &c.— EXAMPLES OF BOILERS AND SPEEDS OF STEAMSHIPS.—NOMINAL AND NORMAL HORSE POWER.

WITH DIAGRAMS FOR ALL BOILER CALCULATIONS AND DRAWINGS OF MANY VARIETIES OF BOILERS.

*** OPINIONS OF THE PRESS.

" The book is one which every mechanical engineer may, with advantage to himself add to his library."—*Industries.*

" Mr. Foley is well fitted to compile such a work. . . . The diagrams are a great feature of the work. . . . Regarding the whole work, it may be very fairly stated that Mr. Foley has produced a volume which will undoubtedly fulfil the desire of the author and become indispensable to all mechanical engineers."—*Marine Engineer.*

" We have carefully examined this work, and pronounce it a most excellent reference book for the use of marine engineers."—*Journal of American Society of Naval Engineers.*

" A veritable monument of industry on the part of Mr. Foley, who has succeeded in producing what is simply invaluable to the engineering profession."—*Steamship.*

Coal and Speed Tables.

A POCKET BOOK OF COAL AND SPEED TABLES, for Engineers and Steam-users. By NELSON FOLEY, Author of " The Mechanical Engineer's Pocket-book." Pocket-size, 3s. 6d. cloth.

" These tables are designed to meet the requirements of every-day use ; they are of sufficient scope for most practical purposes, and may be commended to engineers and users of steam."—*Iron.*

" This pocket-book well merits the attention of the practical engineer. Mr. Foley has compiled a very useful set of tables, the information contained in which is frequently required by engineers, coal consumers and users of steam."—*Iron and Coal Trades Review.*

Steam Engine.

TEXT-BOOK ON THE STEAM ENGINE. With a Supplement on Gas Engines, and PART II. ON HEAT ENGINES. By T. M. GOODEVE, M.A., Barrister-at-Law, Professor of Mechanics at the Royal College of Science, London; Author of "The Principles of Mechanics," "The Elements of Mechanism," &c. Twelfth Edition, Enlarged. With numerous Illustrations. Crown 8vo, 6s. cloth.

' Professor Goodeve has given us a treatise on the steam engine which will bear comparison with anything written by Huxley or Maxwell, and we can award it no higher praise."—*Engineer.*

" Mr. Goodeve's text-book is a work of which every young engineer should possess himself. —*Mining Journal*

Gas Engines.

ON GAS-ENGINES. With Appendix describing a Recent Engine with Tube Igniter. By T. M. GOODEVE, M.A. Crown 8vo, 2s. 6d. cloth. [*Just published.*

"Like all Mr. Goodeve's writings, the present is no exception in point of general excellence It is a valuable little volume."—*Mechanical World.*

Steam Engine Design.

A HANDBOOK ON THE STEAM ENGINE, with especial Reference to Small and Medium-sized Engines. For the Use of Engine-Makers, Mechanical Draughtsmen, Engineering Students and Users of Steam Power. By HERMAN HAEDER, C.E. English Edition, Re-edited by the Author from the Second German Edition, and Translated, with considerable Additions and Alterations, by H. H. P. POWLES, A.M.I.C.E., M.I.M.E. With nearly 1,100 Illustrations. Crown 8vo, 9s. cloth.

"A perfect encyclopædia of the steam engine and its details, and one which must take a permanent place in English drawing-offices and workshops."—*A Foreman Pattern-maker.*

"This is an excellent book, and should be in the hands of all who are interested in the construction and design of medium-sized stationary engines. . . . A careful study of its contents and the arrangement of the sections leads to the conclusion that there is probably no other book like it in this country. The volume aims at showing the results of practical experience, and it certainly may claim a complete achievement of this idea."—*Nature.*

"There can be no question as to its value. We cordially commend it to all concerned in the design and construction of the steam engine."—*Mechanical World.*

Steam Boilers.

A TREATISE ON STEAM BOILERS: Their Strength, Construction, and Economical Working. By ROBERT WILSON, C.E. Fifth Edition 12mo, 6s. cloth.

"The best treatise that has ever been published on steam boilers."—*Engineer.*

"The author shows himself perfect master of his subject, and we heartily recommend all employing steam power to possess themselves of the work."—*Ryland's Iron Trade Circular.*

Boiler Chimneys.

BOILER AND FACTORY CHIMNEYS; Their Draught-Power and Stability. With a Chapter on Lightning Conductors. By ROBERT WILSON, A.I.C.E., Author of "A Treatise on Steam Boilers," &c. Second Edition. Crown 8vo, 3s. 6d. cloth.

"A valuable contribution to the literature of scientific building."—*The Builder.*

Boiler Making.

THE BOILER-MAKER'S READY RECKONER & ASSIST-ANT. With Examples of Practical Geometry and Templating, for the Use of Platers, Smiths and Riveters. By JOHN COURTNEY, Edited by D. K. CLARK, M.I.C.E. Third Edition, 480 pp., with 140 Illusts. Fcap. 8vo, 7s. half-bound.

"No workman or apprentice should be without this book."—*Iron Trade Circular.*

Locomotive Engine Development.

THE LOCOMOTIVE ENGINE AND ITS DEVELOPMENT. A Popular Treatise on the Gradual Improvements made in Railway Engines between 1803 and 1896. By CLEMENT E. STRETTON, C.E., Author of "Safe Railway Working," &c. Fifth Edition, Revised and Enlarged. With 120 Illustrations. Crown 8vo, 3s. 6d. cloth gilt. [*Just published.*

"Students of railway history and all who are interested in the evolution of the modern locomotive will find much to attract and entertain in this volume."—*The Times.*

"The author of this work is well known to the railway world. and no one probably has a better knowledge of the history and development of the locomotive. The volume before us should be of value to all connected with the railway system of this country."—*Nature.*

Estimating for Engineering Work, &c.

ENGINEERING ESTIMATES, COSTS AND ACCOUNTS: A Guide to Commercial Engineering. With numerous Examples of Estimates and Costs of Millwright Work, Miscellaneous Productions, Steam Engines and Steam Boilers; and a Section on the Preparation of Costs Accounts. By A GENERAL MANAGER. Demy 8vo, 12s. cloth.

"This is an excellent and very useful book, covering subject-matter in constant requisition to every factory and workshop. . . . The book is invaluable, not only to the young engineer, but also to the estimate department of every works."—*Builder.*

"We accord the work unqualified praise. The information is given in a plain, straightforward manner, and bears throughout evidence of the intimate practical acquaintance of the author with every phase of commercial engineering."—*Mechanical World.*

Fire Engineering.

FIRES, FIRE-ENGINES, AND FIRE-BRIGADES. With a History of Fire-Engines, their Construction, Use, and Management; Remarks on Fire-Proof Buildings, and the Preservation of Life from Fire; Statistics of the Fire Appliances in English Towns; Foreign Fire Systems; Hints on Fire Brigades, &c. &c. By CHARLES F. T. YOUNG, C.E. With numerous Illustrations. 544 pp., demy 8vo, £1 4s. cloth.
" To those interested in the subject of fires and fire apparatus, we most heartily commend this book. It is the only English work we now have upon the subject."— *Engineering.*

Boilermaking.

PLATING AND BOILERMAKING: A Practical Handbook for Workshop Operations. By JOSEPH G. HORNER, A.M.I.M.E. (Foreman Pattern-Maker), Author of " Pattern Making," &c.   380 pages, with 338 Illustrations. Crown 8vo, 7s. 6d. cloth. | *Just published.*
" The latest production from the pen of this writer is characterised by that evidence of close acquaintance with workshop methods which will render the book exceedingly acceptable to the practical hand. We have no hesitation in commending the work as a serviceable and practical handbook on a subject which has not hitherto received much attention from those qualified to deal with it in a satisfactory manner."—*Mechanical World.*

Engineering Construction.

PATTERN-MAKING: A *Practical Treatise,* embracing the Main Types of Engineering Construction, and including Gearing, both Hand and Machine made, Engine Work, Sheaves and Pulleys, Pipes and Columns, Screws, Machine Parts, Pumps and Cocks, the Moulding of Patterns in Loam and Greensand, &c., together with the methods of Estimating the weight of Castings; to which is added an Appendix of Tables for Workshop Reference. By JOSEPH G. HORNER, A.M.I.M.E. (Foreman Pattern-Maker). Second Edition, thoroughly Revised and much Enlarged. With upwards of 450 Illustrations. Crown 8vo, 7s. 6d. cloth.   [*Just published.*
" A well-written technical guide, evidently written by a man who understands and has practised what he has written about. . . . We cordially recommend it to engineering students, young journeymen, and others desirous of being initiated into the mysteries of pattern-making."—*Builder.*
" More than 450 illustrations help to explain the text, which is, however, always clear and explicit, thus rendering the work an excellent *vade mecum* for the apprentice who desires to become master of his trade."—*English Mechanic.*

Dictionary of Mechanical Engineering Terms.

LOCKWOOD'S DICTIONARY OF TERMS USED IN THE PRACTICE OF MECHANICAL ENGINEERING, embracing those current in the Drawing Office, Pattern Shop, Foundry, Fitting, Turning, Smith's and Boiler Shops, &c. &c. Comprising upwards of 6,000 Definitions. Edited by JOSEPH G. HORNER, A.M.I.M.E. (Foreman Pattern-Maker), Author of " Pattern Making." Second Edition, Revised. Crown 8vo, 7s. 6d. cloth.
" Just the sort of handy dictionary required by the various trades engaged in mechanical engineering. The practical engineering pupil will find the book of great value in his studies, and every foreman engineer and mechanic should have a copy."—*Building News.*
" Not merely a dictionary, but, to a certain extent, also a most valuable guide. It strikes us as a happy idea to combine with a definition of the phrase useful information on the subject of which it treats."—*Machinery Market.*

Mill Gearing.

TOOTHED GEARING: A Practical Handbook for Offices and Workshops. By JOSEPH G. HORNER, A.M.I.M.E. (Foreman Pattern-Maker), Author of " Pattern Making," &c. With 184 Illustrations. Crown 8vo, 6s. cloth.   [*Just published.*

SUMMARY OF CONTENTS.

CHAP. I. PRINCIPLES.—II. FORMATION OF TOOTH PROFILES.—III. PROPORTIONS OF TEETH.—IV. METHODS OF MAKING TOOTH FORMS.—V. INVOLUTE TEETH.— VI. SOME SPECIAL TOOTH FORMS.—VII. BEVEL WHEELS. — VIII. SCREW GEARS. — IX. WORM GEARS.—X. HELICAL WHEELS.—XI. SKEW BEVELS.—XII. VARIABLE AND OTHER GEARS.— XIII. DIAMETRICAL PITCH.—XIV. THE ODONTOGRAPH.— XV. PATTERN GEARS.—XVI. MACHINE MOULDING GEARS.—XVII. MACHINE CUT GEARS.—XVIII. PROPORTION OF WHEELS.

" We must give the book our unqualified praise for its thoroughness of treatment, and we can heartily recommend it to all interested as the most practical book on the subject yet written."—*Mechanical World.*

Stone-working Machinery.

STONE-WORKING MACHINERY, and the Rapid and Economical Conversion of Stone. With Hints on the Arrangement and Management of Stone Works. By M. POWIS BALE, M.I.M.E. With Illusts. Crown 8vo, 9s.

"The book should be in the hands of every mason or student of stone-work."—*Colliery Guardian.*

"A capital handbook for all who manipulate stone for building or ornamental purposes.'—*Machinery Market.*

Pump Construction and Management.

PUMPS AND PUMPING : A Handbook for Pump Users. Being Notes on Selection, Construction and Management. By M. POWIS BALE, M.I.M.E., Author of "Woodworking Machinery," "Saw Mills," &c. Second Edition, Revised. Crown 8vo, 2s. 6d. cloth.

"The matter is set forth as concisely as possible. In fact, condensation rather than diffuseness has been the author's aim throughout; yet he does not seem to have omitted anything likely to be of use."—*Journal of Gas Lighting.*

"Thoroughly practical and simply and clearly written."—*Glasgow Herald.*

Milling Machinery, etc.

MILLING MACHINES AND PROCESSES : A Practical Treatise on Shaping Metals by Rotary Cutters, including Information on Making and Grinding the Cutters. By PAUL N. HASLUCK, Author of "Lathework," "Handybooks for Handicrafts," &c. With upwards of 300 Engravings, including numerous Drawings by the Author. Large crown 8vo, 352 pages, 12s. 6d. cloth.

"A new departure in engineering literature. . . . We can recommend this work to all interested in milling machines ; it is what it professes to be—a practical treatise."—*Engineer.*

"A capital and reliable book, which will no doubt be of considerable service, both to those who are already acquainted with the process as well as to those who contemplate its adoption."
Industries

Turning.

LATHE-WORK : A Practical Treatise on the Tools, Appliances, and Processes employed in the Art of Turning. By PAUL N. HASLUCK. Fifth Edition, Revised and Enlarged Cr. 8vo, 5s. cloth.

"Written by a man who knows, not only how work ought to be done, but who also knows how to do it, and how to convey his knowledge to others. To all turners this book would be valuable."—*Engineering.*

"We can safely recommend the work to young engineers. To the amateur it will simply be invaluable. To the student it will convey a great deal of useful information."—*Engineer.*

Screw-Cutting.

SCREW THREADS : And Methods of Producing Them. With Numerous Tables, and complete directions for using Screw-Cutting Lathes. By PAUL N. HASLUCK, Author of "Lathe-Work," &c. With Seventy-four Illustrations. Third Edition, Revised and Enlarged. Waistcoat-pocket size, 1s. 6d. cloth.

"Full of useful information, hints and practical criticism. Taps, dies and screwing-tools generally are illustrated and their action described."—*Mechanical World.*

"It is a complete compendium of all the details of the screw cutting lathe ; in fact a *multum in parvo* on all the subjects it treats upon."—*Carpenter and Builder.*

Smith's Tables for Mechanics, etc.

TABLES, MEMORANDA, AND CALCULATED RESULTS, FOR MECHANICS, ENGINEERS, ARCHITECTS, BUILDERS, etc. Selected and Arranged by FRANCIS SMITH. Sixth Edition, Revised, including ELECTRICAL TABLES, FORMULÆ, and MEMORANDA. Waistcoat-pocket size, 1s. 6d. limp leather.   [*Just published.*

"It would, perhaps, be as difficult to make a small pocket-book selection of notes and formulæ to suit ALL engineers as it would be to make a universal medicine ; but Mr. Smith's waistcoat-pocket collection may be looked upon as a successful attempt."—*Engineer.*

"The best example we have ever seen of 270 pages of useful matter packed into the dimensions of a card-case."—*Building News.*   "A veritable pocket treasury of knowledge."—*Iron.*

French-English Glossary for Engineers, etc.

A POCKET GLOSSARY of TECHNICAL TERMS : ENGLISH-FRENCH, FRENCH-ENGLISH ; with Tables suitable for the Architectural, Engineering, Manufacturing and Nautical Professions. By JOHN JAMES FLETCHER, Engineer and Surveyor. Second Edition, Revised and Enlarged, 200 pp. Waistcoat-pocket size, 1s. 6d. limp leather.

"It is a very great advantage for readers and correspondents in France and England to have so large a number of the words relating to engineering and manufacturers collected in a lilliputian volume. The little book will be useful both to students and travellers."—*Architect.*

"The glossary of terms is very complete, and many of the tables are new and well arranged. We cordially commend the book."—*Mechanical World.*

Year-Book of Engineering Formulæ, &c.

THE ENGINEER'S YEAR-BOOK FOR 1896. Comprising Formulæ, Rules, Tables, Data and Memoranda in Civil, Mechanical, Electrical, Marine and Mine Engineering. By H. R. KEMPE, A.M. Inst.C.E., M.I.E.E., Technical Officer of the Engineer-in-Chief's Office, General Post Office, London, Author of "A Handbook of Electrical Testing," "The Electrical Engineer's Pocket-Book," &c. With 800 Illustrations, specially Engraved for the work. Crown 8vo, 670 pages, 8s. leather. [*Just published.*

"Represents an enormous quantity of work and forms a desirable book of reference."—*The Engineer.*

"The book is distinctly in advance of most similar publications in this country."—*Engineering.*

"This valuable and well-designed book of reference meets the demands of all descriptions of engineers."—*Saturday Review.*

"Teems with up-to-date information in every branch of engineering and construction."—*Building News.*

"The needs of the engineering profession could hardly be supplied in a more admirable, complete and convenient form. To say that it more than sustains all comparisons is praise of the highest sort, and that may justly be said of it."—*Mining Journal.*

"There is certainly room for the new comer, which supplies explanations and directions, as well as formulæ and tables. It deserves to become one of the most successful of the technical annuals."—*Architect.*

"Brings together with great skill all the technical information which an engineer has to use day by day. It is in every way admirably equipped, and is sure to prove successful."—*Scotsman.*

"The up-to-dateness of Mr. Kempe's compilation is a quality that will not be lost on the busy people for whom the work is intended."—*Glasgow Herald.*

Portable Engines.

THE PORTABLE ENGINE; ITS CONSTRUCTION AND MANAGEMENT. A Practical Manual for Owners and Users of Steam Engines generally. By WILLIAM DYSON WANSBROUGH. With 90 Illustrations. Crown 8vo, 3s. 6d. cloth.

"This is a work of value to those who use steam machinery. . . . Should be read by everyone who has a steam engine, on a farm or elsewhere."—*Mark Lane Express.*

"We cordially commend this work to buyers and owners of steam engines, and to those who have to do with their construction or use."—*Timber Trades Journal.*

"Such a general knowledge of the steam engine as Mr. Wansbrough furnishes to the reader should be acquired by all intelligent owners and others who use the steam engine."—*Building News.*

"An excellent text-book of this useful form of engine. 'The Hints to Purchasers' contain a good deal of commonsense and practical wisdom."—*English Mechanic.*

Iron and Steel.

"IRON AND STEEL": A Work for the Forge, Foundry, Factory, and Office. Containing ready, useful, and trustworthy Information for Ironmasters and their Stock-takers; Managers of Bar, Rail, Plate, and Sheet Rolling Mills; Iron and Metal Founders; Iron Ship and Bridge Builders; Mechanical, Mining, and Consulting Engineers; Architects, Contractors, Builders, and Professional Draughtsmen. By CHARLES HOARE, Author of "The Slide Rule," &c. Ninth Edition. 32mo, 6s. leather.

"For comprehensiveness the book has not its equal."—*Iron.*

"One of the best of the pocket books."—*English Mechanic.*

"We cordially recommend this book to those engaged in considering the details of all kinds of iron and steel works."—*Naval Science.*

Elementary Mechanics.

CONDENSED MECHANICS. A Selection of Formulæ, Rules, Tables, and Data for the Use of Engineering Students, Science Classes, &c. In Accordance with the Requirements of the Science and Art Department. By W. G. CRAWFORD HUGHES, A.M.I.C.E. Crown 8vo, 2s 6d. cloth.

"The book is well fitted for those who are either confronted with practical problems in their work, or are preparing for examination and wish to refresh their knowledge by going through their formulæ again."—*Marine Engineer.*

"It is well arranged, and meets the wants of those for whom it is intended."—*Railway News.*

Steam.

THE SAFE USE OF STEAM. Containing Rules for Unprofessional Steam-users. By an ENGINEER. Seventh Edition. Sewed, 6d.

"If steam-users would but learn this little book by heart, boiler explosions would become sensations by their rarity."—*English Mechanic.*

Warming.

HEATING BY HOT WATER; with Information and Suggestions on the best Methods of Heating Public, Private and Horticultural Buildings. By WALTER JONES. Second Edition. With 96 Illustrations. Crown 8vo, 2s. 6d. net.

"We confidently recommend all interested in heating by hot water to secure a copy of this valuable little treatise."—*The Plumber and Decorator.*

CIVIL ENGINEERING, SURVEYING, etc.

Light Railways.

LIGHT RAILWAYS FOR THE UNITED KINGDOM, INDIA, AND THE COLONIES : A Practical Handbook setting forth the Principles on which Light Railways should be Constructed, Worked and Financed ; and detailing the cost of Construction, Equipment, Revenue, and Working Expences of Local Railways already established in the above-mentioned Countries, and in Belgium, France, Switzerland, &c. By JOHN CHARLES MACKAY, F.G.S., A.M.Inst.C E. Illustrated with 40 Photographic Plates and other Diagrams. Medium 8vo, 15s. cloth. [*Just published.*

"Exactly what has been long wanted, and sure to have a wide sale."—*Railway News.*

Water Supply and Water-Works.

THE WATER SUPPLY OF TOWNS AND THE CON-STRUCTION OF WATER-WORKS: A Practical Treatise for the Use of Engineers and Students of Engineering. By W. K. BURTON, A.M.Inst.C E., Professor of Sanitary Engineering in the Imperial University, Tokyo, Japan, and Consulting Engineer to the Tokyo Water-works. With an Appendix on The Effects of Earthquakes on Waterworks, by JOHN MILNE, F.R.S., Pro-fessor of Mining in the Imperial University of Japan. With numerous Plates and Illustrations. Super-royal 8vo, 25s. buckram. [*Just published.*

" The whole art of waterworks construction is dealt with in a clear and comprehensive fashion in this handsome volume. . . . Mr. Burton's practical treatise shows in all its sections the fruit of independent study and individual experience. It is largely based upon his own practice in the branch of engineering of which it treats, and with such a basis a treatise can scarcely fail to be suggestive and useful."—*Saturday Review.*

" Professor Burton's book is sure of a warm welcome among engineers. It is written in clear and vigorous language and forms an exhaustive treatise on a branch of engineering the claims of which it would be difficult to over-estimate."—*Scotsman.*

" The subjects seem to us to be ably discussed, with a practical aim to meet the requirements of all its probable readers. The volume is well got up, and the illustrations are excellent."
The Lancet.

Water Supply of Cities and Towns.

A COMPREHENSIVE TREATISE on the WATER-SUPPLY OF CITIES AND TOWNS. By WILLIAM HUMBER, A-M.Inst.C.E., and M. Inst. M.E., Author of "Cast and Wrought Iron Bridge Construction," &c. &c. Illustrated with 50 Double Plates, 1 Single Plate, Coloured Frontispiece, and upwards of 250 Woodcuts, and containing 400 pages of Text. Imp. 4to, £6 6s. elegantly and substantially half-bound in morocco.

" The most systematic and valuable work upon water supply hitherto produced in English or in any other language. . . . Mr. Humber's work is characterised almost throughout by an exhaustiveness much more distinctive of French and German than of English technical treatises. —*Engineer.*

" We can congratulate Mr. Humber on having been able to give so large an amount of information on a subject so important as the water supply of cities and towns. The plates, fifty in number, are mostly drawings of executed works, and alone would have commanded the attention of every engineer whose practice may lie in this branch of the profession."—*Builder.*

Water Supply.

RURAL WATER SUPPLY : A Practical Handbook on the Supply of Water and Construction of Waterworks for small Country Districts. By ALLAN GREENWELL, A.M.I.C.E., and W. T. CURRY, A.M.I.C.E., F.G.S. With Illustrations. Crown 8vo, 5s. cloth. [*Just published.*

" We conscientiously recommend it as a very useful book for those concerned in obtaining water for small districts, giving a great deal of practical information in a small compass."—*Builder.*

" The volume contains valuable information upon all matters connected with water supply. . . . It is full of details on points which are continually before waterworks' engineers."—*Nature.*

Hydraulic Tables.

HYDRAULIC TABLES, CO-EFFICIENTS, and FORMULÆ for finding the Discharge of Water from Orifices, Notches, Weirs, Pipes, and Rivers. By JOHN NEVILLE, Civil Engineer, M.R.I.A. Third Ed., carefully Revised, with considerable Additions. Numerous Illusts. Cr. 8vo, 14s. cloth.

" Alike valuable to students and engineers in practice ; its study will prevent the annoyance of avoidable failures, and assist them to select the readiest means of successfully carrying out any given work connected with hydraulic engineering."—*Mining Journal.*

" It is, of all English books on the subject, the one nearest to completeness, . .—*Architect*

Hydraulics.

HYDRAULIC MANUAL. Consisting of Working Tables and Explanatory Text. Intended as a Guide in Hydraulic Calculations and Field Operations. By LOWIS D'A. JACKSON, Author of "Aid to Survey Practice," "Modern Metrology," &c. Fourth Edition, Enlarged. Large cr. 8vo, 16s. cl.

"The author has had a wide experience in hydraulic engineering and has been a careful observer of the facts which have come under his notice, and from the great mass of material at his command he has constructed a manual which may be accepted as a trustworthy guide to this branch of the engineer's profession. We can heartily recommend this volume to all who desire to be acquainted with the latest development of this important subject."—*Engineering.*
"The standard-work in this department of mechanics."—*Scotsman.*
"The most useful feature of this work is its freedom from what is superannuated, and its thorough adoption of recent experiments; the text is, in fact, in great part a short account of the great modern experiments."—*Nature.*

Water Storage, Conveyance, and Utilisation.

WATER ENGINEERING : A Practical Treatise on the Measurement, Storage, Conveyance, and Utilisation of Water for the Supply of Towns, for Mill Power, and for other Purposes. By CHARLES SLAGG, A.M.Inst.C.E., Author of "Sanitary Work in the Smaller Towns, and in Villages," &c. Second Edition. With numerous Illustrations. Crown 8vo, 7s. 6d. cloth.

"As a small practical treatise on the water supply of towns, and on some applications of water-power, the work is in many respects excellent."—*Engineering.*
"The author has collated the results deduced from the experiments of the most eminent authorities, and has presented them in a compact and practical form, accompanied by very clear and detailed explanations. . . . The application of water as a motive power is treated very carefully and exhaustively."—*Builder.*
"For anyone who desires to begin the study of hydraulics with a consideration of the practical applications of the science there is no better guide."—*Architect.*

Drainage.

ON THE DRAINAGE OF LANDS, TOWNS, AND BUILDINGS. By G. D. DEMPSEY, C.E., Author of "The Practical Railway Engineer," &c. Revised, with large Additions on RECENT PRACTICE IN DRAINAGE ENGINEERING, by D. KINNEAR CLARK, M.Inst.C.E. Author of "Tramways: Their Construction and Working," "A Manual of Rules. Tables, and Data for Mechanical Engineers," &c. Third Edition. Small crown 8vo, 4s. 6d. cloth. [*Just published.*

"The new matter added to Mr. Dempsey's excellent work is characterised by the comprehensive grasp and accuracy of detail for which the name of Mr. D. K. Clark is a sufficient voucher."—*Athenaeum.*
"As a work on recent practice in drainage engineering, the book is to be commended to all who are making that branch of engineering science their special study."—*Iron.*
"A comprehensive manual on drainage engineering, and a useful introduction to the student."—*Building News.*

River Engineering.

RIVER BARS: The Causes of their Formation, and their Treatment by "Induced Tidal Scour;" with a Description of the Successful Reduction by this Method of the Bar at Dublin. By I. J. MANN, Assist. Eng. to the Dublin Port and Docks Board. Royal 8vo, 7s. 6d. cloth.

"We recommend all interested in harbour works—and, indeed, those concerned in the improvements of rivers generally—to read Mr. Mann's interesting work on the treatment of river bars."—*Engineer.*

Tramways and their Working.

TRAMWAYS : THEIR CONSTRUCTION AND WORKING. Embracing a Comprehensive History of the System; with an exhaustive Analysis of the various Modes of Traction, including Horse-Power, Steam, Cable Traction, Electric Traction, &c.; a Description of the Varieties of Rolling Stock; and ample Details of Cost and Working Expenses. New Edition, Thoroughly Revised, and Including the Progress recently made in Tramway Construction, &c. &c. By D. KINNEAR CLARK, M.Inst.C.E. With numerous Illustrations and Folding Plates. In One Volume, 8vo, 780 pages, price 28s., bound in buckram. [*Just published.*

"All interested in tramways must refer to it, as all railway engineers have turned to the author's work 'Railway Machinery.'"—*Engineer.*
"An exhaustive and practical work on tramways, in which the history of this kind of locomotion and a description and cost of the various modes of laying tramways, are to be found."—*Building News.*
"The best form of rails, the best mode of construction, and the best mechanical appliances are so fairly indicated in the work under review, that any engineer about to construct a tramway will be enabled at once to obtain the practical information which will be of most service to him.' —*Athenaeum.*

Student's Text-Book on Surveying.

PRACTICAL SURVEYING : A Text-Book for Students preparing for Examination or for Survey-work in the Colonies. By GEORGE
W. USILL, A.M.I.C.E., Author of "The Statistics of the Water Supply of
Great Britain." With Four Lithographic Plates and upwards of 330 Illustrations. Fourth Edition, Revised, including Tables of Natural Sines, Tangents, Secants, &c. Crown 8vo, 7s. 6d. cloth ; or, on THIN PAPER, bound in
limp leather, gi t edges, rounded corners, for pocket use, 12s. 6d.

"The best forms of instruments are described as to their construction, uses and modes of
employment, and there are innumerable hints on work and equipment such as the author, in his
experience as surveyor, draughtsman, and teacher, has found necessary, and which the student
in his inexperience will find most serviceable."—*Engineer.*

"The latest treatise in the English language on surveying, and we have no hesitation in saying that the student will find it a better guide than any of its predecessors . . . .
Deserves to be recognised as the first book which should be put in the hands of a pupil of Civil
Engineering, and every gentleman of education who sets out for the Colonies would find it well to
have a copy."—*Architect.*

Survey Practice.

*AID TO SURVEY PRACTICE, for Reference in Surveying, Level-
ling, and Setting-out ; and in Route Surveys of Travellers by Land and Sea.*
With Tables, Illustrations, and Records. By LOWIS D'A. JACKSON,
A.M.I.C.E., Author of "Hydraulic Manual," "Modern Metrology," &c.
Second Edition, Enlarged. Large crown 8vo, 12s. 6d. cloth.

"A valuable *vade-mecum* for the surveyor. We can recommend this book as containing an
admirable supplement to the teaching of the accomplished surveyor."—*Athenæum.*

"As a text-book we should advise all surveyors to place it in their libraries, and study well the
matured instructions afforded in its pages."—*Colliery Guardian.*

"The author brings to his work a fortunate union of theory and practical experience which,
aided by a clear and lucid style of writing, renders the book a very useful one."—*Builder.*

Surveying, Land and Marine.

LAND AND MARINE SURVEYING, in Reference to the Preparation of Plans for Roads and Railways; Canals, Rivers, Towns' Water
Supplies; Docks and Harbours. With Description and Use of Surveying
Instruments. By W. D. HASKOLL, C.E., Author of "Bridge and Viaduct Construction," &c. Second Edition, Revised, with Additions. Large cr. 8vo, 9s. cl.

"This book must prove of great value to the student. We have no hesitation in recommending it, feeling assured that it will more than repay a careful study."—*Mechanical World.*

"A most useful and well arranged book. We can strongly recommend it as a carefully-written
and valuable text-book. It enjoys a well-deserved repute among surveyors."—*Builder.*

"This volume cannot fail to prove of the utmost practical utility. It may be safely recommended
to all students who aspire to become clean and expert surveyors."—*Mining Journal.*

Field-Book for Engineers.

*THE ENGINEER'S, MINING SURVEYOR'S, AND CON-
TRACTOR'S FIELD-BOOK.* Consisting of a Series of Tables, with Rules,
Explanations of Systems, and use of Theodolite for Traverse Surveying and
Plotting the Work with minute accuracy by means of Straight Edge and Set
Square only ; Levelling with the Theodolite, Casting-out and Reducing
Levels to Datum, and Plotting Sections in the ordinary manner; setting-out
Curves with the Theodolite by Tangential Angles and Multiples, with Right
and Left-hand Readings of the Instrument: Setting-out Curves without
Theodolite, on the System of Tangential Angles by sets of Tangents and Off-
sets ; and Earthwork Tables to 80 feet deep, calculated for every 6 inches in
depth. By W. D. HASKOLL, C.E. Fourth Edition. Crown 8vo, 12s. cloth.

"The book is very handy; the separate tables of sines and tangents to every minute will make
t useful for many other purposes, the genuine traverse tables existing all the same."—*Athenæum.*

"Every person engaged in engineering field operations will estimate the importance of such a
work and the amount of valuable time which will be saved by reference to a set of reliable tables
prepared with the accuracy and fulness of those given in this volume."—*Railway News.*

Levelling.

*A TREATISE ON THE PRINCIPLES AND PRACTICE OF
LEVELLING.* Showing its Application to purposes of Railway and Civil
Engineering, in the Construction of Roads; with Mr. TELFORD's Rules for the
same. By FREDERICK W. SIMMS, F.G.S., M.Inst.C.E. Seventh Edition, with
the addition of LAW's Practical Examples for Setting-out Railway Curves, and
TRAUTWINE's Field Practice of Laying-out Circular Curves. With 7 Plates
and numerous Woodcuts. 8vo, 8s. 6d. cloth. ******* TRAUTWINE on Curves
may be had separate, 5s.

"The text-book on levelling in most of our engineering schools and colleges. . . " . The
publishers have rendered a substantial service to the profession, especially to the younger members,
by bringing out the present edition of Mr. Simms's useful book "—*Engineer.*

Trigonometrical Surveying.

AN OUTLINE OF THE METHOD OF CONDUCTING A TRIGONOMETRICAL SURVEY, for the Formation of Geographical and Topographical Maps and Plans, Military Reconnaissance, Levelling, &c., with Useful Problems, Formulæ, and Tables. By Lieut.-General FROME, R.E. Fourth Edition, Revised and partly Re-written by Major General Sir CHARLES WARREN, G.C.M.G., R.E. With 19 Plates and 115 Woodcuts. Royal 8vo, 16s. cloth.

"The simple fact that a fourth edition has been called for is the best testi mony to its merits No words of praise from us can strengthen the position so well and so steadily maintained by this work. Sir Charles Warren has revised the entire work, and made such additions as were necessary to bring every portion of the contents up to the present date."—*Broad Arrow.*

Curves, Tables for Setting-out.

TABLES OF TANGENTIAL ANGLES AND MULTIPLES for Setting-out Curves from 5 to 200 Radius. By ALEXANDER BEAZELEY, M.Inst.C.E. Fourth Edition. Printed on 48 Cards, and sold in a cloth box, waistcoat-pocket size, 3s. 6d.

"Each table is printed on a small card, which, being placed on the theodolite, leaves the hand free to manipulate the instrument—no small advantage as regards the rapidity of work."—*Engineer.*
"Very handy; a man may know that all his day's work must fall on two of these cards, which he puts into his own card-case, and leaves the rest behind."—*Athenæum.*

Earthwork.

HANDY GENERAL EARTHWORK TABLES. Giving the Contents in Cub:c Yards of Centre and Slopes of Cuttings and Embankments from 3 inches to 80 feet in Depth or Height, for use with either 66 feet Chain or 100 feet Chain. By J. H. WATSON BUCK, M.Inst.C.E. On a Sheet mounted in cloth case, 3s. 6d.   [*Just published.*

Earthwork.

EARTHWORK TABLES. Showing the Contents in Cubic Yards of Embankments, Cuttings, &c., of Heights or Depths up to an average of 80 feet. By JOSEPH BROADBENT, C.E., and FRANCIS CAMPIN, C.E. Crown 8vo, 5s. cloth.

"The way in which accuracy is attained, by a simple division of each cross section into three elements, two in which are constant and one variable, is ingenious."—*Athenæum.*

Earthwork, Measurement of.

A MANUAL ON EARTHWORK. By ALEX. J. S. GRAHAM, C.E. With numerous Diagrams. Second Edition. 18mo, 2s. 6d. cloth.

"A great amount of practical information, very admirably arranged, and available for rough e:timates, as well as for the more exact calculations required in the engineer's and contractor's offices."—*Artisan.*

Tunnelling.

PRACTICAL TUNNELLING. Explaining in detail the Setting-out of the works, Shaft-sinking and Heading-driving, Ranging the Lines and Levelling underground, Sub-Excavating, Timbering, and the Construction of the Brickwork of Tunnels, with the amount of Labour required for, and the Cost of, the various portions of the work. By FREDERICK W. SIMMS, M.Inst. C.E. Fourth Edition, Revised and Further Extended including the Most Recent (1895) Examples of Sub aqueous and other Tunnels, by D. KINNEAR CLARK, M.Inst. C.E. Imperial 8vo, with 34 Folding Plates and other Illustrations, £2 2s. cloth.   [*Just published.*

"The estimation in which Mr. Simms's book on tunnelling has been held for over thirty years cannot be more truly expressed than in the words of the late Prof. Rankine :—' The best source of information on the subject of tunnels is Mr. F.W. Simms's work on Practical Tunnelling.'"—*Architect.*
"It has been regarded from the first as a text-book of the subject. . . . Mr. Clark has added immensely to the value of the book."—*Engineer.*

Tunnel Shafts.

THE CONSTRUCTION OF LARGE TUNNEL SHAFTS : A Practical and Theoretical Essay. By J. H. WATSON BUCK, M.Inst.C.E., Resident Engineer, London and North-Western Railway. Illustrated with Folding Plates. Royal 8vo, 12s. cloth.

"Many of the methods given are of extreme practical value to the mason ; and the observation s on the form of arch, the rules for ordering the stone, and the construction of the templates will be found of considerable use. We commend the book to the engineering profession."—*Building News.*
"Will be regarded by civil engineers as of the utmost value, and calculated to save much time and obviate many mistakes."—*Colliery Guardian.*

Oblique Bridges.

A PRACTICAL AND THEORETICAL ESSAY ON OBLIQUE BRIDGES. With 13 large Plates. By the late GEORGE WATSON BUCK, M.I.C.E. Fourth Edition, revised by his Son, J. H. WATSON BUCK, M.I.C.E.; and with the addition of Description to Diagrams for Facilitating the Construction of Oblique Bridges, by W. H. BARLOW, M.I.C.E. Roy. 8vo, 12s. cl.

"The standard text-book for all engineers regarding skew arches is Mr. Buck's treatise, and it would be impossible to consult a better."—*Engineer.*

"Mr. Buck's treatise is recognised as a standard text-book, and his treatment has divested the subject of many of the intricacies supposed to belong to it. As a guide to the engineer and architect, on a confessedly difficult subject, Mr. Buck's work is unsurpassed."—*Building News.*

Cast and Wrought Iron Bridge Construction.

A COMPLETE AND PRACTICAL TREATISE ON CAST AND WROUGHT IRON BRIDGE CONSTRUCTION, including Iron Foundations. In Three Parts—Theoretical, Practical, and Descriptive. By WILLIAM HUMBER, A.M.Inst.C.E., and M.Inst.M.E. Third Edition, Revised and much improved, with 115 Double Plates (20 of which now first appear in this edition), and numerous Additions to the Text. In Two Vols., imp. 4to, £6 16s. 6d. half-bound in morocco.

"A very valuable contribution to the standard literature of civil engineering. In addition to elevations, plans and sections, large scale details are given which very much enhance the instructive worth of those illustrations."—*Civil Engineer and Architect's Journal.*

Iron Bridges.

IRON BRIDGES OF MODERATE SPAN : Their Construction and Erection. By HAMILTON WELDON PENDRED, late Inspector of Ironwork to the Salford Corporation. With 40 Illustrations. 12mo, 2s. cloth.

"Students and engineers should obtain this book for constant and practical use."—*Colliery Guardian.*

Oblique Arches.

A PRACTICAL TREATISE ON THE CONSTRUCTION OF OBLIQUE ARCHES. By JOHN HART. Third Edition, with Plates. Imperial 8vo, 8s. cloth.

Statics, Graphic and Analytic.

GRAPHIC AND ANALYTIC STATICS, in their Practical Application to the Treatment of Stresses in Roofs, Solid Girders, Lattice, Bowstring and Suspension Bridges, Braced Iron Arches and Piers, and other Frameworks. By R. HUDSON GRAHAM, C.E. Containing Diagrams and Plates to Scale. With numerous Examples, many taken from existing Structures. Specially arranged for Class-work in Colleges and Universities. Second Edition, Revised and Enlarged. 8vo, 16s. cloth.

"Mr. Graham's book will find a place wherever graphic and analytic statics are used or studied."—*Engineer.*

"The work is excellent from a practical point of view, and has evidently been prepared with much care. The directions for working are ample, and are illustrated by an abundance of well-selected examples. It is an excellent text-book for the practical draughtsman."—*Athenæum.*

Girders, Strength of.

GRAPHIC TABLE FOR FACILITATING THE COMPUTATION OF THE WEIGHTS OF WROUGHT IRON AND STEEL GIRDERS, etc., for Parliamentary and other Estimates. By J. H. WATSON BUCK, M.Inst.C.E. On a Sheet, 2s. 6d.

Strains, Calculation of.

A HANDY BOOK FOR THE CALCULATION OF STRAINS IN GIRDERS AND SIMILAR STRUCTURES, AND THEIR STRENGTH. Consisting of Formulæ and Corresponding Diagrams, with numerous details for Practical Application, &c. By WILLIAM HUMBER, A.M.Inst.C.E., &c. Fifth Edition. Crown 8vo, nearly 100 Woodcuts and 3 Plates, 7s. 6d. cloth.

"The formulæ are neatly expressed, and the diagrams good."—*Athenæum.*

"We heartily commend this really *handy* book to our engineer and architect readers."—*English Mechanic.*

Trusses.

TRUSSES OF WOOD AND IRON. *Practical Applications of Science in Determining the Stresses, Breaking Weights, Safe Loads, Scantlings, and Details of Construction*, with Complete Working Drawings. By WILLIAM GRIFFITHS, Surveyor, Assistant Master, Tranmere School of Science and Art. Oblong 8vo, 4s. 6d. cloth.

" This handy little book enters so minutely into every detail connected with the construction of roof trusses, that no student need be ignorant of these matters."—*Practical Engineer.*

Strains in Ironwork.

THE STRAINS ON STRUCTURES OF IRONWORK; with Practical Remarks on Iron Construction. By F. W. SHEILDS, M.Inst.C.E, Second Edition, with 5 Plates. Royal 8vo, 5s. cloth.

" The student cannot find a better little book on this subject."—*Engineer.*

Barlow's Strength of Materials, enlarged by Humber.

A TREATISE ON THE STRENGTH OF MATERIALS: with Rules for Application in Architecture, the Construction of Suspension Bridges, Railways, &c. By PETER BARLOW, F.R.S. A New Edition, Revised by his Sons, P. W. BARLOW, F.R.S., and W. H. BARLOW, F.R.S.; to which are added, Experiments by HODGKINSON, FAIRBAIRN, and KIRKALDY; and Formulæ for Calculating Girders, &c. Arranged and Edited by WM. HUMBER, A-M.Inst.C.E. Demy 8vo, 400 pp., with 19 large Plates and numerous Woodcuts, 18s. cloth.

" Valuable alike to the student, tyro, and the experienced practitioner, it will always rank in future, as it has hitherto done, as the standard treatise on that particular subject."—*Engineer.*
" There is no greater authority than Barlow."—*Building News.*
" As a scientific work of the first class, it deserves a foremost place on the bookshelves of every civil engineer and practical mechanic."—*English Mechanic.*

Cast Iron and other Metals, Strength of.

A PRACTICAL ESSAY ON THE STRENGTH OF CAST IRON AND OTHER METALS. By THOMAS TREDGOLD, C.E. Fifth Edition, including HODGKINSON's Experimental Researches. 8vo, 12s. cloth.

Railway Working.

SAFE RAILWAY WORKING. *A Treatise on Railway Accidents: Their Cause and Prevention; with a Description of Modern Appliances and Systems.* By CLEMENT E. STRETTON, C.E. With Illustrations and Coloured Plates. Third Edition, Enlarged. Crown 8vo, 3s. 6d.

" A book for the engineer, the directors, the managers; and, in short, all who wish for information on railway matters will find a perfect encyclopædia in 'Safe Railway Working.'"—*Railway Review.*
" We commend the remarks on railway signalling to all railway managers, especially where a uniform code and practice is advocated.'—*Herepath's Railway Journal.*
" The author may be congratulated on having collected, in a very convenient form, much valuable information on the principal questions affecting the safe working of railways."—*Railway Engineer.*

Heat, Expansion by.

EXPANSION OF STRUCTURES BY HEAT. By JOHN KEILY, C.E., late of the Indian Public Works and Victorian Railway Departments. Crown 8vo, 3s. 6d. cloth.

SUMMARY OF CONTENTS.

Section   I. FORMULAS AND DATA.
Section II. METAL BARS.
Section III. SIMPLE FRAMES.
Section IV. COMPLEX FRAMES AND PLATES.
Section V. THERMAL CONDUCTIVITY.
Section VI. MECHANICAL FORCE OF HEAT.
Section VII. WORK OF EXPANSION AND CONTRACTION.
Section VIII. SUSPENSION BRIDGES.
Section IX. MASONRY STRUCTURES.

" The aim the author has set before him, viz., to show the effects of heat upon metallic and other structures, is a laudable one, for this is a branch of physics upon which the engineer or architect can find but little reliable and comprehensive data in books."—*Builder.*
" Whoever is concerned to know the effect of changes of temperature on such structures as suspension bridges and the like, could not do better than consult Mr. Keily's valuable and handy exposition of the geometrical principles involved in these changes."—*Scotsman.*

Field Fortification.

A TREATISE ON FIELD FORTIFICATION, THE ATTACK OF FORTRESSES, MILITARY MINING, AND RECONNOITRING. By Colonel I. S. MACAULAY, late Professor of Fortification in the R.M.A., Woolwich. Sixth Edition. Crown 8vo, with separate Atlas of 12 Plates, 12s. cloth.

MR. HUMBER'S GREAT WORK ON MODERN ENGINEERING.

Complete in Four Volumes, imperial 4to, price £12 12s., half-morocco. Each Volume sold separately as follows:—

A RECORD OF THE PROGRESS OF MODERN ENGINEER-ING. FIRST SERIES.

Comprising Civil, Mechanical, Marine, Hydraulic, Railway, Bridge, and other Engineering Works, &c. By WILLIAM HUMBER, A-M.Inst.C.E., &c. Imp. 4to, with 36 Double Plates, drawn to a large scale, Photographic Portrait of John Hawkshaw, C.E., F.R.S., &c., and copious descriptive Letterpress, Specifications, &c., £3 3s. half-morocco.

List of the Plates and Diagrams.

Victoria Station and Roof, L. B. & S. C. R. (8 plates); Southport Pier (2 plates); Victoria Station and Roof, L. C. & D. and G. W. R. (6 plates); Roof of Cremorne Music Hall; Bridge over G. N. Railway; Roof of Station, Dutch Rhenish Rail (2 plates); Bridge over the | Thames, West London Extension Railway (5 plates); Armour Plates: Suspension Bridge, Thames (4 plates); The Allen Engine; Suspension Bridge, Avon (3 plates); Underground Railway (3 plates).

"Handsomely lithographed and printed. It will find favour with many who desire to preserve in a permanent form copies of the plans and specifications prepared for the guidance of the contractors for many important engineering works."—*Engineer.*

HUMBER'S PROGRESS OF MODERN ENGINEERING. SECOND SERIES.

Imp. 4to, with 36 Double Plates, Photographic Portrait of Robert Stephenson, C.E., M.P., F.R.S., &c., and copious descriptive Letterpress, Specifications, &c., £3 3s. half-morocco.

List of the Plates and Diagrams.

Birkenhead Docks, Low Water Basin (15 plates); Charing Cross Station Roof, C. C. Railway (3 plates); Digswell Viaduct, Great Northern Railway; Robbery Wood Viaduct, Great Northern Railway; Iron Permanent Way; Clydach Viaduct, Merthyr, Tredegar, | and Abergavenny Railway; Ebbw Viaduct, Merthyr, Tredegar, and Abergavenny Railway; College Wood Viaduct, Cornwall Railway; Dublin Winter Palace Roof (3 plates); Bridge over the Thames, L. C. & D. Railway (6 plates); Albert Harbour, Greenock (4 plates).

"Mr. Humber has done the profession good and true service, by the fine collection of examples he has here brought before the profession and the public."—*Practical Mechanic's Journal.*

HUMBER'S PROGRESS OF MODERN ENGINEERING. THIRD SERIES.

Imp. 4to, with 40 Double Plates, Photographic Portrait of J. R. M'Clean, late Pres. Inst. C.E., and copious descriptive Letterpress, Specifications, &c., £3 3s. half-morocco.

List of the Plates and Diagrams.

MAIN DRAINAGE, METROPOLIS.—*North Side.*—Map showing Interception of Sewers; Middle Level Sewer (2 plates); Outfall Sewer, Bridge over River Lea (3 plates); Outfall Sewer, Bridge over Marsh Lane, North Woolwich Railway, and Bow and Barking Railway Junction; Outfall Sewer, Bridge over Bow and Barking Railway (3 plates); Outfall Sewer, Bridge over East London Waterworks' Feeder (2 plates); Outfall Sewer, Reservoir (2 plates); Outfall Sewer, Tumbling Bay and Outlet; Outfall Sewer, Penstocks. *South Side.*—Outfall Sewer, Bermondsey Branch (2 plates); Outfall | Sewer, Reservoir and Outlet (4 plates); Outfall Sewer, Filth Hoist; Sections of Sewers (North and South Sides). THAMES EMBANKMENT.—Section of River Wall; Steamboat Pier, Westminster (2 plates); Landing Stairs between Charing Cross and Waterloo Bridges; York Gate (2 plates); Overflow and Outlet at Savoy Street Sewer (3 plates); Steamboat Pier, Waterloo Bridge (3 plates); Junction of Sewers, Plans and Sections; Gullies, Plans and Sections; Rolling Stock; Granite and Iron Forts.

"The drawings have a constantly increasing value, and whoever desires to possess clear representations of the two great works carried out by our Metropolitan Board will obtain Mr. Humber's volume."—*Engineer.*

HUMBER'S PROGRESS OF MODERN ENGINEERING. FOURTH SERIES.

Imp. 4to, with 36 Double Plates, Photographic Portrait of John Fowler, late Pres. Inst. C.E., and copious descriptive Letterpress, Specifications, &c., £3 3s. half-morocco.

List of the Plates and Diagrams.

Abbey Mills Pumping Station, Main Drainage, Metropolis (4 plates); Barrow Docks (5 plates); Manquis Viaduct, Santiago and Valparaiso Railway (2 plates); Adam's Locomotive, St. Helen's Canal Railway (2 plates); Cannon Street Station Roof, Charing Cross Railway (3 plates); Road Bridge over the River Moka (2 plates); Telegraph Apparatus for | Mesopotamia; Viaduct over the River Wye, Midland Railway (3 plates); St. Germans Viaduct, Cornwall Railway (2 plates); Wrought-Iron Cylinder for Diving Bell; Millwall Docks (6 plates); Milroy's Patent Excavator; Metropolitan District Railway (6 plates); Harbours, Ports, and Breakwaters (3 plates).

"We gladly welcome another year's issue of this valuable publication from the able pen of Mr. Humber. The accuracy and general excellence of this work are well known, while its usefulness in giving the measurements and details of some of the latest examples of engineering, as carried out by the most eminent men in the profession, cannot be too highly prized."—*Artisan*

THE POPULAR WORKS OF MICHAEL REYNOLDS

("THE ENGINE DRIVER'S FRIEND ").

Locomotive-Engine Driving.

LOCOMOTIVE-ENGINE DRIVING : A Practical Manual for Engineers in charge of Locomotive Engines. By MICHAEL REYNOLDS, Member of the Society of Engineers, formerly Locomotive Inspector L. B. and S. C. R. Ninth Edition. Including a KEY TO THE LOCOMOTIVE ENGINE. With Illustrations and Portrait of Author. Crown 8vo, 4s. 6d. cloth.

"Mr. Reynolds has supplied a want, and has supplied it well. We can confidently recommend the book, not only to the practical driver, but to everyone who takes an interest in the performance of locomotive engines."—*The Engineer.*

"Mr. Reynolds has opened a new chapter in the literature of the day. Of the practical utility of this admirable treatise, we have to speak in terms of warm commendation."—*Athenæum.*

"Evidently the work of one who knows his subject thoroughly."—*Railway Service Gazette.*

"Were the cautions and rules given in the book to become part of the every-day working of our engine-drivers, we might have fewer distressing accidents to deplore."—*Scotsman.*

Stationary Engine Driving.

STATIONARY ENGINE DRIVING : A Practical Manual for Engineers in charge of Stationary Engines. By MICHAEL REYNOLDS. Fifth Edition. Enlarged. With Plates and Woodcuts. Crown 8vo, 4s. 6d. cloth.

"The author is thoroughly acquainted with his subjects, and his advice on the various points treated is clear and practical. . . . He has produced a manual which is an exceedingly useful one for the class for whom it is specially intended."—*Engineering.*

"Our author leaves no stone unturned. He is determined that his readers shall not only know something about the stationary engine, but all about it."—*Engineer.*

"An engineman who has mastered the contents of Mr. Reynolds's book will require but little actua experience with boilers and engines before he can be trusted to look after them."—*English Mechanic.*

The Engineer, Fireman, and Engine-Boy.

THE MODEL LOCOMOTIVE ENGINEER, FIREMAN, and ENGINE-BOY. Comprising a Historical Notice of the Pioneer Locomotive Engines and their Inventors. By MICHAEL REYNOLDS. Second Edition, with Revised Appendix. With numerous Illustrations and Portrait of George Stephenson. Crown 8vo, 4s. 6d. cloth.   [*Just published.*

" From the technical knowledge of the author it will appeal to the railway man of to-day more forcibly than anything written by Dr. Smiles. . . . The volume contains information of a technical kind, and facts that every driver should be familiar with."—*English Mechanic.*

"We should be glad to see this book in the possession of everyone in the kingdom who has ever laid, or is to lay, hands on a locomotive engine."—*Iron.*

Continuous Railway Brakes.

CONTINUOUS RAILWAY BRAKES : A Practical Treatise on the several Systems in Use in the United Kingdom ; their Construction and Performance. With copious Illustrations and numerous Tables. By MICHAEL REYNOLDS. Large crown 8vo, 9s. cloth.

" A popular explanation of the different brakes. It will be of great assistance in forming public opinion, and will be studied with benefit by those who take an interest in the brake."—*English Mechanic.*

"Written with sufficient technical detail to enable the principle and relative connection of the various parts of each particular brake to be readily grasped."—*Mechanical World.*

Engine-Driving Life.

ENGINE-DRIVING LIFE : Stirring Adventures and Incidents in the Lives of Locomotive-Engine Drivers. By MICHAEL REYNOLDS. Third and Cheaper Edition. Crown 8vo, 1s. 6d. cloth.

"From first to last perfectly fascinating. Wilkie Collins's most thrilling conceptions are thrown into the shade by true incidents, endless in their variety, related in every page."—*North British Mail.*

"Anyone who wishes to get a real insight into railway life cannot do better than read ' Engine-Driving Life' for himself ; and if he once take it up he will find that the author's enthusiasm and real love of the engine-driving profession will carry him on till he has read every page."—*Saturday Review.*

Pocket Companion for Enginemen.

THE ENGINEMAN'S POCKET COMPANION AND PRAC-TICAL EDUCATOR FOR ENGINEMEN, BOILER ATTENDANTS, AND MECHANICS. By MICHAEL REYNOLDS. With Forty-five Illustrations and numerous Diagrams. Third Edition, Revised. Royal 18mo, 3s. 6d., strongly bound for pocket wear.

This admirable work is well suited to accomplish its object, being the honest workmanship of a competent engineer."—*Glasgow Herald.*

''A most meritorious work, giving in a succinct and practical form all the information an engine-minder desirous of mastering the scientific principles of his daily calling would require."—*The Miller.*

" A boon to those who are striving to become efficient mechanics."—*Daily Chronicle.*

MARINE ENGINEERING, SHIPBUILDING, NAVIGATION, etc.

Pocket-Book for Naval Architects and Shipbuilders,
THE NAVAL ARCHITECT'S AND SHIPBUILDER'S
POCKET-BOOK *of Formulæ, Rules, and Tables, and* MARINE ENGINEER'S
AND SURVEYOR'S *Handy Book of Reference.* By CLEMENT MACKROW,
Member of the Institution of Naval Architects, Naval Draughtsman. Sixth
Edition, Revised. 700 pages, with upwards of 300 Illustrations. Fcap., 12s.'6d.
strongly bound in leather. [*Just published.*]

SUMMARY OF CONTENTS.

SIGNS AND SYMBOLS, DECIMAL FRAC-
TIONS.— TRIGONOMETRY. — PRACTICAL
GEOMETRY. — MENSURATION. — CEN-
TRES AND MOMENTS OF FIGURES.—
MOMENTS OF INERTIA AND RADII OF
GYRATION. — ALGEBRAICAL EXPRES-
SIONS FOR SIMPSON'S RULES.—ME-
CHANICAL PRINCIPLES. — CENTRE OF
GRAVITY.—LAWS OF MOTION.—DIS-
PLACEMENT, CENTRE OF BUOYANCY.—
CENTRE OF GRAVITY OF SHIP'S HULL.
—STABILITY CURVES AND METACEN-
TRES.—SEA AND SHALLOW-WATER
WAVES.—ROLLING OF SHIPS.—PRO-
PULSION AND RESISTANCE OF VESSELS.
—SPEED TRIALS.—SAILING, CENTRE
OF EFFORT.—DISTANCES DOWN RIVERS,
COAST LINES.—STEERING AND RUD-
DERS OF VESSELS.—LAUNCHING CAL-
CULATIONS AND VELOCITIES.—WEIGHT
OF MATERIAL AND GEAR.—GUN PAR-
TICULARS AND WEIGHT.—STANDARD
GAUGES.—RIVETED JOINTS AND RIVET-
ING.—STRENGTH AND TESTS OF MATE-
RIALS. — BINDING AND SHEARING
STRESSES, ETC.—STRENGTH OF SHAFT-
ING, PILLARS, WHEELS, ETC. — HY-
DRAULIC DATA, ETC.—CONIC SECTIONS,
CATENARIAN CURVES.—MECHANICAL
POWERS, WORK. — BOARD OF TRADE
REGULATIONS FOR BOILERS AND EN-
GINES. — BOARD OF TRADE REGULA-
TIONS FOR SHIPS.—LLOYD'S RULES

FOR BOILERS.—LLOYD'S WEIGHT OF
CHAINS.—LLOYD'S SCANTLINGS FOR
SHIPS.—DATA OF ENGINES AND VES-
SELS. - SHIPS' FITTINGS AND TESTS.—
SEASONING PRESERVING TIMBER.—
MEASUREMENT OF TIMBER.—ALLOYS,
PAINTS, VARNISHES. — DATA FOR
STOWAGE. — ADMIRALTY TRANSPORT
REGULATIONS. — RULES FOR HORSE-
POWER, SCREW PROPELLERS, ETC.—
PERCENTAGES FOR BUTT STRAPS, ETC.
—PARTICULARS OF YACHTS.—MASTING
AND RIGGING VESSELS.—DISTANCES
OF FOREIGN PORTS. — TONNAGE
TABLES. — VOCABULARY OF FRENCH
AND ENGLISH TERMS. — ENGLISH
WEIGHTS AND MEASURES —FOREIGN
WEIGHTS AND MEASURES.—DECIMAL
EQUIVALENTS. — FOREIGN MONEY.—
DISCOUNT AND WAGE TABLES.—USE-
FUL NUMBERS AND READY RECKONERS
—TABLES OF CIRCULAR MEASURES.—
TABLES OF AREAS OF AND CIRCUM-
FERENCES OF CIRCLES.—TABLES OF
AREAS OF SEGMENTS OF CIRCLES.—
TABLES OF SQUARES AND CUBES AND
ROOTS OF NUMBERS. — TABLES OF
LOGARITHMS OF NUMBERS.—TABLES
OF HYPERBOLIC LOGARITHMS.—TABLES
OF NATURAL SINES, TANGENTS, ETC.—
TABLES OF LOGARITHMIC SINES,
TANGENTS, ETC.

" In these days of advanced knowledge a work like this is of the greatest value. It contains a
vast amount of information. We hesitatingly say that it is the most valuable compilation for its
specific purpose that has ever been printed. No naval architect, engineer, surveyor, or seaman,
wood or iron shipbuilder, can afford to be without this work."—*Nautical Magazine.*

"Should be used by all who are engaged in the construction or designs of vessels. . . . Will
be found to contain the most useful tables and formulæ required by shipbuilders, carefully collected
from the best authorities, and put together in a popular and simple form."—*Engineer.*

"The professional shipbuilder has now, in a convenient and accessible form, reliable data for
solving many of the numerous problems that present themselves in the course of his work."—*Iron.*

"There is no doubt that a pocket-book of this description must be a necessity in the ship-
building trade. . . . The volume contains a mass of useful information clearly expressed and
presented in a handy form."—*Marine Engineer.*

Marine Engineering.

MARINE ENGINES AND STEAM VESSELS *(A Treatise
on).* By ROBERT MURRAY, C.E. Eighth Edition, thoroughly Revised, with
considerable Additions by the Author and by GEORGE CARLISLE, C.E.,
Senior Surveyor to the Board of Trade at Liverpool. 12mo, 5s. cloth boards.

"Well adapted to give the young steamship engineer or marine engine and boiler maker a
general introduction into his practical work."—*Mechanical World.*

"We feel sure that this thoroughly revised edition will continue to be as popular in the future
as it has been in the past, as, for its size, it contains more useful information than any similar
treatise."—*Industries.*

"As a compendious and useful guide to engineers of our mercantile and royal naval services,
we should say it cannot be surpassed."—*Building News.*

" The information given is both sound and sensible, and well qualified to direct young sea-
going hands on the straight road to the extra chief's certificate. . . . Most useful to surveyors,
inspectors, draughtsmen, and young engineers."—*Glasgow Herald.*

English-French Dictionary of Sea Terms.

TECHNICAL DICTIONARY OF SEA TERMS, PHRASES
AND WORDS USED IN THE ENGLISH & FRENCH LANGUAGES.
(English-French, French-English). For the Use of Seamen, Engineers,
Pilots, Ship-builders, Ship-owners and Ship-brokers. Compiled by W.
PIRRIE, late of the African Steamship Company. Fcap. 8vo, 5s. cloth limp.
[*Just published.*

"This volume will be highly appreciated by seamen, engineers, pilots, shipbuilders and ship-owners. It will be found wonderfully accurate and complete."—*Scotsman.*
"A very useful dictionary, which has l ng been wanted by French and English engineers,
masters, officers and others."—*Shipping World.*

Pocket-Book for Marine Engineers.

A POCKET-BOOK OF USEFUL TABLES AND FOR-
MULÆ FOR MARINE ENGINEERS. By FRANK PROCTOR, A.I.N.A.
Third Edition. Royal 32mo, leather, gilt edges, with strap, 4s.
"We recommend it to our readers as going far to supply a long-felt want."—*Naval Science.*
"A most useful companion to all marine engineers."—*United Service Gazette.*

Introduction to Marine Engineering.

ELEMENTARY ENGINEERING : *A Manual for Young Marine
Engineers and Apprentices.* In the Form of Questions and Answers on
Metals, Alloys, Strength of Materials, Construction and Management of
Marine Engines and Boilers, Geometry, &c. &c. With an Appendix of Useful
Tables. By JOHN SHERREN BREWER, Government Marine Surveyor, Hong-
kong. Third Edition. Small crown 8vo, 1s. 6d. cloth.
"Contains much valuable information for the class for whom it is intended, especially in the
chapters on the management of boilers and engines."—*Nautical Magazine.*
"A useful introduction to the more elaborate text-books."—*Scotsman.*
"To a student who has the requisite desire and resolve to attain a thorough knowledge, Mr.
Brewer offers decidedly useful help."—*Athenaum.*

Navigation.

PRACTICAL NAVIGATION. Consisting of THE SAILOR'S
SEA-BOOK, by JAMES GREENWOOD and W. H. ROSSER ; together with the
requisite Mathematical and Nautical Tables for the Working of the Problems,
by HENRY LAW, C.E., and Professor J. R. YOUNG. Illustrated. 12mo, 7s.
strongly half-bound.

Sailmaking.

THE ART AND SCIENCE OF SAILMAKING. By SAMUEL
B. SADLER, Practical Sailmaker, late in the employment of Messrs. Ratsey
and Lapthorne, of Cowes and Gosport. With Plates and other Illustrations.
Small 4to, 12s. 6d. cloth.
"This work is very ably written, and is illustrated by diagrams and carefully-worked calculations. The work should be in the hands of every sailmaker, whether employer or employed, as it
cannot fail to assist them in the pursuit of their important avocations."—*Isle of Wight Herald.*
"This extremely practical work gives a complete education in all the branches of the manufacture cutting out, roping, seaming, and goring. It is copiously illustrated, and will form a first-
rate text-book and guide."—*Portsmouth Times.*
"The author of this work has rendered a distinct service to all interested in the art of sail-
making. The subject of which he treats is a congenial one. Mr. Sadler is a practical sailmaker
and has devoted years of careful observation and study to the subject ; and the results of the
experience thus gained he has set forth in the volume before us."—*Steamship.*

Chain Cables.

CHAIN CABLES AND CHAINS. Comprising Sizes and
Curves of Links, Studs, &c., Iron for Cables and Chains, Chain Cable and
Chain Making, Forming and Welding Links, Strength of Cables and Chains,
Certificates for Cables, Marking Cables, Prices of Chain Cables and Chains,
Historical Notes, Acts of Parliament, Statutory Tests, Charges for Testing,
List of Manufacturers of Cables, &c. &c. By THOMAS W. TRAILL, F.E.R.N.,
M. Inst. C.E., Engineer Surveyor in Chief, Board of Trade, Inspector of
Chain Cable and Anchor Proving Establishments, and General Superin-
tendent, Lloyd's Committee on Proving Establishments. With numerous
Tables, Illustrations and Lithographic Drawings. Folio, £2 2s. cloth.
"It contains a vast amount of valuable information. Nothing seems to be wanting to make it
a complete and standard work of reference on the subject."—*Nautical Magazine.*

MINING AND METALLURGY.

Mining Machinery.

MACHINERY FOR METALLIFEROUS MINES: A Practical Treatise for Mining Engineers, Metallurgists, and Managers of Mines. By E. HENRY DAVIES, M.E., F.G.S. Crown 8vo, 580 pp., with upwards of 300 Illustrations, 12s. 6d. cloth. [*Just published.*

"Mr. Davies, in this handsome volume, has done the advanced student and the manager of mines good service. Almost every kind of machinery in actual use is carefully described, and the woodcuts and plates are good."—*Athenæum.*

"From cover to cover the work exhibits all the same characteristics which excite the confidence and attract the attention of the student as he peruses the first page. The work may safely be recommended. By its publication the literature connected with the industry will be enriched, and the reputation of its author enhanced."—*Mining Journal.*

"Mr. Davies has endeavoured to bring before his readers the best of everything in modern mining appliances. His work carries internal evidence of the author's impartiality, and this constitutes one of the great merits of the book. Throughout his work the criticisms are based on his own or other reliable experience.'—*Iron and Steel Trades' Journal.*

"The work deals with nearly every class of machinery or apparatus likely to be met with or required in connection with metalliferous mining, and is one which we have every confidence in recommending."—*Practical Engineer.*

Metalliferous Minerals and Mining.

A TREATISE ON METALLIFEROUS MINERALS AND MINING. By D. C. DAVIES, F.G.S., Mining Engineer, &c., Author of "A Treatise on Slate and Slate Quarrying." Fifth Edition, thoroughly Revised and much Enlarged, by his Son, E. HENRY DAVIES, M.E., F.G.S. With about 150 Illustrations. Crown 8vo, 12s. 6d. cloth.

"Neither the practical miner nor the general reader interested in mines can have a better book for his companion and his guide."—*Mining Journal.* [*Mining World.*

"We are doing our readers a service in calling their attention to this valuable work."

"A book that will not only be useful to the geologist, the practical miner, and the metallurgist but also very interesting to the general public."—*Iron.*

"As a history of the present state of mining throughout the world this book has a real value and it supplies an actual want."—*Athenæum.*

Earthy Minerals and Mining.

A TREATISE ON EARTHY & OTHER MINERALS AND MINING. By D. C. DAVIES, F.G.S., Author of "Metalliferous Minerals,' &c. Third Edition, revised and Enlarged, by his Son, E. HENRY DAVIES M.E., F.G.S. With about 100 Illustrations. Crown 8vo, 12s. 6d. cloth.

"We do not remember to have met with any English work on mining matters that contains the same amount of information packed in equally convenient form."—*Academy.*

"We should be inclined to rank it as among the very best of the handy technical and trades manuals which have recently appeared."—*British Quarterly Review.*

Metalliferous Mining in the United Kingdom.

BRITISH MINING: A Treatise on the History, Discovery, Practical Development, and Future Prospects of Metalliferous Mines in the United Kingdom. By ROBERT HUNT, F.R.S., Editor of "Ure's Dictionary of Arts, Manufactures, and Mines," &c. Upwards of 950 pp., with 230 Illustrations. Second Edition, Revised. Super-royal 8vo, £2 2s. cloth.

"One of the most valuable works of reference of modern times. Mr. Hunt, as Keeper of Mining Records of the United Kingdom, has had opportunities for such a task not enjoyed by anyone else and has evidently made the most of them. . . . The language and style adopted are good, and the treatment of the various subjects laborious, conscientious, and scientific."—*Engineering.*

"The book is, in fact, a treasure-house of statistical information on mining subjects, and we know of no other work embodying so great a mass of matter of this kind. Were this the only merit of Mr. Hunt s volume, it would be sufficient to render it indispensable in the library of everyone interested in the development of the mining and metallurgical industries of this country. —*Athenæum.*

"A mass of information not elsewhere available, and of the greatest value to those who may be interested in our great mineral industries."—*Engineer.*

Underground Pumping Machinery.

MINE DRAINAGE. Being a Complete and Practical Treatise on Direct-Acting Underground Steam Pumping Machinery, with a Description of a large number of the best known Engines, their General Utility and the Special Sphere of their Action, the Mode of their Application, and their merits compared with other forms of Pumping Machinery. By STEPHEN MICHELL. 8vo, 15s. cloth.

"Will be highly esteemed by colliery owners and lessees, mining engineers, and students generally who require to be acquainted with the best means of securing the drainage of mines. It is a most valuable work, and stands almost alone in the literature of steam pumping machinery."—*Colliery Guardian.*

"Much valuable information is given, so that the book is thoroughly worthy of an extensive circulation amongst practical men and purchasers of machinery."—*Mining Journal.*

Prospecting for Gold and other Metals.

THE PROSPECTOR'S HANDBOOK : A Guide for the Prospector and Traveller in Search of Metal-Bearing or other Valuable Minerals. By J. W. ANDERSON, M.A. (Camb.), F.R.G.S., Author of "Fiji and New Caledonia," Sixth Edition, thoroughly Revised and much Enlarged. Small crown 8vo, 3s. 6d. cloth ; or, 4s. 6d. leather, pocket-book form, with tuck.

[*Just published.*

"Will supply a much felt want, especially among Colonists, in whose way are so often thrown many mineralogical specimens the value of which it is difficult to determine."—*Engineer.*
"How to find commercial minerals, and how to identify them when they are found, are the leading points to which attention is directed. The author has managed to pack as much practical detail into his pages as would supply material for a book three times its size."—*Mining Journal.*

Mining Notes and Formulæ.

NOTES AND FORMULÆ FOR MINING STUDENTS. By JOHN HERMAN MERIVALE, M.A., Certificated Colliery Manager, Professor of Mining in the Durham College of Science, Newcastle-upon-Tyne. Third Edition, Revised and Enlarged. Small crown 8vo, 2s. 6d. cloth.

"Invaluable to anyone who is working up for an examination on mining subjects."—*Iron and Coal Trades Review.*
"The author has done his work in an exceedingly creditable manner, and has produced a book that will be of service to students, and those who are practically engaged in mining operations."—*Engineer.*

Handybook for Miners.

THE MINER'S HANDBOOK : A Handy Book of Reference on the Subjects of Mineral Deposits, Mining Operations, Ore Dressing, &c. For the Use of Students and others interested in Mining matters. Compiled by JOHN MILNE, F.R.S., Professor of Mining in the Imperial University of Japan. Revised Edition. Fcap. 8vo, 7s. 6d. leather. [*Just published.*

"Professor Milne's handbook is sure to be received with favour by all connected with mining, and will be extremely popular among students."—*Athenæum.*

Miners' and Metallurgists' Pocket-Book.

A POCKET-BOOK FOR MINERS AND METALLURGISTS. Comprising Rules, Formulæ, Tables, and Notes, for Use in Field and Office Work. By F. DANVERS POWER, F.G.S., M.E. Fcap. 8vo, 9s. leather.

"This excellent book is an admirable example of its kind, and ought to find a large sale amongst English-speaking prospectors and mining engineers."—*Engineering.*
"A useful *vade-mecum* containing a mass of rules, formulæ, tables, and various other information, necessary for daily reference."—*Iron.*

Mineral Surveying and Valuing.

THE MINERAL SURVEYOR AND VALUER'S COMPLETE GUIDE, *comprising a Treatise on Improved Mining Surveying and the Valuation of Mining Properties, with New Traverse Tables.* By WM. LINTERN. Third Edition, Enlarged. 12mo, 4s. cloth.

"A valuable and thoroughly trustworthy guide."—*Iron and Coal Trades Review.*

Asbestos and its Uses.

ASBESTOS : *Its Properties, Occurrence, and Uses.* With some Account of the Mines of Italy and Canada. By ROBERT H. JONES. With Eight Collotype Plates and other Illustrations. Crown 8vo, 12s. 6d. cloth.

"An interesting and invaluable work."—*Colliery Guardian.*

Iron, Metallurgy of.

METALLURGY OF IRON. Containing History of Iron Manufacture, Methods of Assay, and Analyses of Iron Ores, Processes of Manufacture of Iron and Steel, &c. By H. BAUERMAN, F.G.S., A.R.S.M. With numerous Illustrations. Sixth Edition, Enlarged. 12mo, 5s. 6d. cloth.

"Carefully written, it has the merit of brevity and conciseness, as to less important points; while all material matters are very fully and thoroughly entered into."—*Standard.*

Slate Quarrying, &c.

SLATE AND SLATE QUARRYING, Scientific, Practical and Commercial. By D. C. DAVIES, F.G.S., Mining Engineer, &c. With numerous Illustrations and Folding Plates. Third Edition, 12mo, 3s. cloth.

"One of the best and best-balanced treatises on a special subject that we have met with."—*Engineer.*

Colliery Management.

THE COLLIERY MANAGER'S HANDBOOK: A Comprehensive Treatise on the Laying-out and Working of Collieries. Designed as a Book of Reference for Colliery Managers, and for the Use of Coal-Mining Students preparing for First-class Certificates. By CALEB PAMELY, Mining Engineer and Surveyor; Member of the North of England Institute of Mining and Mechanical Engineers; and Member of the South Wales Institute of Mining Engineers. With nearly 700 Plans, Diagrams, and other Illustrations. Third Edition, Revised and Enlarged. Medium 8vo. about 900 pages. Price £1 5s. strongly bound. [*Just published.*

SUMMARY OF CONTENTS.

GEOLOGY. — SEARCH FOR COAL.— MINERAL LEASES AND OTHER HOLDINGS.—SHAFT SINKING.—FITTING UP THE SHAFT AND SURFACE ARRANGEMENTS.—STEAM BOILERS AND THEIR FITTINGS.—TIMBERING AND WALLING. —NARROW WORK AND METHODS OF WORKING.—UNDERGROUND CONVEYANCE.—DRAINAGE.—THE GASES MET WITH IN MINES; VENTILATION.—ON THE FRICTION OF AIR IN MINES.—

THE PRIESTMAN OIL ENGINE; PETROLEUM AND NATURAL GAS—SURVEYING AND PLANNING.—LIGHTING; SAFETY LAMPS AND FIRE DAMP DETECTORS.— SUNDRY AND INCIDENTAL OPERATIONS AND APPLIANCES.—COLLIERY EXPLOSIONS. — MISCELLANEOUS QUESTIONS AND ANSWERS.

Appendix: SUMMARY OF REPORT OF H.M. COMMISSIONERS ON ACCIDENTS IN MINES.

"There can be no doubt that it is the best book on coal-mining."—J. T. ROBSON, Esq., *H.M.'s Inspector of Mines, South Wales District.*

" Mr. Pamely's work is eminently suited to the purpose for which it is intended—being clear, interesting, exhaustive, rich in detail, and up to date, giving descriptions of the very latest machines in every department. . . . A mining engineer could scarcely go wrong who followed this work."—*Colliery Guardian.*

" This is the most complete 'all round' work on coal-mining published in the English language. . . . No library of coal-mining books is complete without it."—*Colliery Engineer* (Scranton, Pa., U.S.A.).

" Mr. Pamely's work is in all respects worthy of our admiration. No person in any responsible position connected with mines should be without a copy."—*Westminster Review.*

Coal and Iron.

THE COAL AND IRON INDUSTRIES OF THE UNITED KINGDOM. Comprising a Description of the Coal Fields, and of the Principal Seams of Coal, with Returns of their Produce and its Distribution, and Analyses of Special Varieties. Also an Account of the occurrence of Iron Ores in Veins or Seams; Analyses of each Variety; and a History of the Rise and Progress of Pig Iron Manufacture. By RICHARD MEADE, Assistant Keeper of Mining Records. With Maps. 8vo, £1 8s. cloth.

" The book is one which must find a place on the shelves of all interested in coal and iron production, and in the iron, steel, and other metallurgical industries."—*Engineer.*

"Of this book we may unreservedly say that it is the best of its class which we have ever met. . . . A book of reference which no one engaged in the iron or coal trades should omit from his library."—*Iron and Coal Trades Review.*

Coal Mining.

COAL AND COAL MINING: A Rudimentary Treatise on. By the late Sir WARINGTON W. SMYTH, M.A., F.R.S., Chief Inspector of the Mines of the Crown. Seventh Edition, Revised and Enlarged. With numerous Illustrations. 12mo, 4s. cloth boards.

'As an outline is given of every known coal-field in this and other countries, as well as of the principal methods of working, the book will doubtless interest a very large number of readers."— *Mining Journal.*

Subterraneous Surveying.

SUBTERRANEOUS SURVEYING, Elementary and Practical Treatise on, with and without the Magnetic Needle. By THOMAS FENWICK, Surveyor of Mines, and THOMAS BAKER, C.E. Illust. 12mo, 3s. cloth boards.

Granite Quarrying.

GRANITES AND OUR GRANITE INDUSTRIES. By GEORGE F. HARRIS, F.G.S., Membre de la Société Belge de Géologie, Lecturer on Economic Geology at the Birkbeck Institution, &c. With Illustrations. Crown 8vo, 2s. 6d. cloth.

" A clearly and well-written manual on the granite industry."—*Scotsman.*

" An interesting work, which will be deservedly esteemed."—*Colliery Guardian.*

" An exceedingly interesting and valuable monograph on a subject which has hitherto received unaccountably little attention in the shape of systematic literary treatment."—*Scottish Leader.*

Gold, Metallurgy of.

THE METALLURGY OF GOLD: A Practical Treatise on the Metallurgical Treatment of Gold-bearing Ores. Including the Processes of Concentration, Chlorination and Extraction by Cyanide, and the Assaying, Melting, and Refining of Gold. By M. EISSLER, Mining Engineer and Metallurgical Chemist, formerly Assistant Assayer of the U.S. Mint, San Francisco. Fourth Edition, Enlarged. With about 250 Illustrations and numerous Folding Plates and Working Drawings. 8vo, 16s. cloth.   [*Just published.*

"This book thoroughly deserves its title of a 'Practical Treatise.' The whole process of gold milling, from the breaking of the quartz to the assay of the bullion is described in clear and orderly narrative and with much, but not too much, fulness of detail."—*Saturday Review.*

"The work is a storehouse of information and valuable data, and we strongly recommend it to all professional men engaged in the gold-mining industry."—*Mining Journal.*

Gold Extraction.

THE CYANIDE PROCESS OF GOLD EXTRACTION: and its Practical Application on the Witwatersrand Gold Fields in South Africa. By M. EISSLER, M.E., Mem. Inst. Mining and Metallurgy, Author of "The Metallurgy cf Gold," &c. With Diagrams and Working Drawings. Large crown 8vo, 7s. 6d. cloth.   [*Just published.*

"This book is just what was needed to acquaint mining men with the actual working of a process which is not only the most popular, but is, as a general rule, the most successful for the extraction of gold from tailings."—*Mining Journal.*

"The work will prove invaluable to all interested in gold mining, whether metallurgists or as investors."—*Chemical News.*

Silver, Metallurgy of.

THE METALLURGY OF SILVER: A Practical Treatise on the Amalgamation, Roasting, and Lixiviation of Silver Ores. Including the Assaying, Melting and Refining, of Silver Bullion. By M. EISSLER, Author of "The Metallurgy of Gold," &c. Third Edition. With 150 Illustrations. Crown 8vo, 10s. 6d. cloth.   [*Just published.*

"A practical treatise, and a technical work which we are convinced will supply a long-felt want amongst practical men, and at the same time be of value to students and others indirectly connected with the industries."—*Mining Journal.*

"From first to last the book is thoroughly sound and reliable."—*Colliery Guardian.*

"For chemists, practical miners, assayers, and investors alike, we do not know of any work on the subject so handy and yet so comprehensive."—*Glasgow Herald.*

Lead, Metallurgy of.

THE METALLURGY OF ARGENTIFEROUS LEAD: A Practical Treatise on the Smelting of Silver-Lead Ores and the Refining of Lead Bullion. Including Reports on various Smelting Establishments and Descriptions of Modern Smelting Furnaces and Plants in Europe and America. By M. EISSLER, M.E., Author of "The Metallurgy of Gold," &c. Crown 8vo, 400 pp., with 183 Illustrations, 12s. 6d. cloth.

"The numerous metallurgical processes, which are fully and extensively treated of, embrace all the stages experienced in the passage of the lead from the various natural states to its issue from the refinery as an article of commerce."—*Practical Engineer.*

"The present volume fully maintains the reputation of the author. Those who wish to obtain a thorough insight into the present state of this industry cannot do better than read this volume, and all mining engineers cannot fail to find many useful hints and suggestions in it."—*Industries.*

"It is most carefully written and illustrated with capital drawings and diagrams. In fact, it is the work of an expert for experts, by whom it will be prized as an indispensable text-book."—*Bristol Mercury.*

Iron Mining.

THE IRON ORES OF GREAT BRITAIN AND IRELAND: Their Mode of Occurrence, Age, and Origin, and the Methods of Searching for and Working them, with a Notice of some of the Iron Ores of Spain. By J. D. KENDALL, F.G.S., Mining Engineer. Crown 8vo, 16s. cloth.

"The author has a thorough practical knowledge of his subject, and has supplemented a careful study of the available literature by unpublished information derived from his own observations. The result is a very useful volume which cannot fail to be of value to all interested in the iron industry of the country."—*Industries.*

"Mr. Kendall is a great authority on this subject and writes from personal observation."—*Colliery Guardian.*

"Mr. Kendall's book is thoroughly well done. In it there are the outlines of the history of ore mining in every centre and there is everything that we want to know as to the character of the ores of each district, their commercial value and the cost of working them"—*Iron and Steel Trades Journal.*

ELECTRICITY, ELECTRICAL ENGINEERING, etc.

Dynamo Management.

THE MANAGEMENT OF DYNAMOS : A Handybook of
Theory and Practice for the Use of Mechanics, Engineers, Students and
others in Charge of Dynamos. By G. W. LUMMIS PATERSON. With numerous Illustrations. Crown 8vo, 3s. 6d. cloth.   [*Just published*.
" An example which deserves to be taken as a model by other authors. The subject is treated
in a manner which any intelligent man who is fit to be entrusted with charge of an engine should
be able to understand. It is a useful book to all who make, tend or employ electric machinery."—
Architect.
" A most satisfactory book from a practical point of view. We strongly commend it to the
attention of every electrical engineering student."—*Daily Chronicle.*

Electrical Engineering.

THE ELECTRICAL ENGINEER'S POCKET-BOOK OF
MODERN RULES, FORMULÆ, TABLES, AND DATA. By H. R.
KEMPE, M.Inst.E.E., A.M.Inst.C.E., Technical Officer, Postal Telegraphs,
Author of " A Handbook of Electrical Testing," &c. Second Edition,
thoroughly Revised, with Additions. Royal 32mo, oblong, 5s. leather.
" There is very little in the shape of formulæ or data which the electrician is likely to want
in a hurry which cannot be found in its pages."—*Practical Engineer.*
" A very useful book of reference for daily use in practical electrical engineering and its
various applications to the industries of the present day."—*Iron.*
" It is the best book of its kind."—*Electrical Engineer.*
" Well arranged and compact. The 'Electrical Engineer's Pocket-Book' is a good one."—
Electrician.   [*Review.*
" Strongly recommended to those engaged in the various electrical industries."—*Electrical*

Electric Lighting.

ELECTRIC LIGHT FITTING : A Handbook for Working
Electrical Engineers, embodying Practical Notes on Installation Manage-
ment. By JOHN W. URQUHART, Electrician, Author of " Electric Light," &c.
With numerous Illustrations. Second Edition, Revised, with Additional
Chapters. Crown 8vo, 5s. cloth.
" This volume deals with what may be termed the mechanics of electric lighting, and is
addressed to men who are already engaged in the work or are training for it. The work traverses
a great deal of ground, and may be read as a sequel to the same author's useful work on ' Electric
Light.' "—*Electrician.*
" The book is well worth the perusal of the workmen for whom it is written."—*Electrical
Review.*
" We have read this book with a good deal of pleasure. We believe that the book will be of
use to practical workmen, who will not be alarmed by finding mathematical formulæ which they
are unable to understand."—*Electrical Plant.*

Electric Light.

ELECTRIC LIGHT : Its Production and Use. Embodying Plain
Directions for the Treatment of Dynamo-Electric Machines, Batteries,
Accumulators, and Electric Lamps. By J. W. URQUHART, C.E., Author of
" Electric Light Fitting," " Electrop'ating," &c. Fifth Edition, carefully
Revised, with Large Additions and 145 Illustrations. Crown 8vo, 7s. 6d. cloth.
" The whole ground of electric lighting is more or less covered and explained in a very clear
and concise manner."—*Electrical Review.*
" Contains a good deal of very interesting information, especially in the parts where the
author gives dimensions and working costs."—*Electrical Engineer.*
" A miniature *vade-mecum* of the salient facts connected with the science of electric light-
ing."—*Electrician.*
" You cannot have a better book than ' Electric Light,' by Urquhart."— *Engineer.*
" The book is by far the best that we have yet met with on the subject."—*Athenæum.*

Construction of Dynamos.

DYNAMO CONSTRUCTION : A Practical Handbook for the Use
of Engineer Constructors and Electricians-in-Charge. Embracing Frame-
work Building, Field Magnet and Armature Winding and Grouping, Com-
pounding, &c. With Examples of leading English, American, and Continental Dynamos and Motors. By J. W. URQUHART, Author of "Electric
Light," "Electric Light Fitting," &c. Second Edition, Revised and Enlarged. With 114 Illustrations. Crown 8vo, 7s. 6d. cloth. [*Just published.*
" Mr. Urquhart's book is the first one which deals with these matters in such a way that the
engineering student can understand them. The book is very readable, and the author leads his
readers up to difficult subjects by reasonably simple tests."—*Engineering Review.*
" The author deals with his subject in a style so popular as to make his volume a handbook of
great practical value to engineer constructors and electricians in charge."— *Scotsman.*
" ' Dynamo Construction' more than sustains the high character of the author's previous
publications. It is sure to be widely read by the large and rapidly increasing number of practical
electricians."—*Glasgow Herald.*

New Dictionary of Electricity.

THE STANDARD ELECTRICAL DICTIONARY. A Popu-
lar Dictionary of Words and Terms Used in the Practice of Electrical Engineering. Containing upwards of 3,000 Definitions. By T. O'CONNOR SLOANE,
A.M., Ph.D., Author of "The Arithmetic of Electricity," &c. Crown 8vo,
630 pp., 350 Illustrations, 7s. 6d. cloth. [Just published.

"The work has many attractive features in it, and is beyond doubt, a well put together and
useful publication. The amount of ground covered may be gathered from the fact that in the
index about 5,000 references will be found. The inclusion of such comparatively modern words
as 'impedence,' 'reluctance,' &c., shows that the author has desired to be up to date, and indeed
there are other indications of carefulness of compilation. The work is one which does the author
great credit and it should prove of great value, especially to students."—Electrical Review.

Very complete and contains a large amount of useful information."—Industries.

"An encyclopædia of electrical science in the compass of a dictionary. The information
given is sound and clear. The book is well printed, well illustrated, and well up to date, and may
be confidently recommended."—Builder.

"The volume is excellently printed and illustrated, and should form part of the library of
every one who is connected with electrical matters."—Hardware Trade Journal.

Electric Lighting of Ships.

ELECTRIC SHIP-LIGHTING : A Handbook on the Practical
Fitting and Running of Ship's Electrical Plant. For the Use of Shipowners
and Builders, Marine Electricians, and Sea-going Engineers-in-Charge. By
J. W. URQUHART, C.E. With 88 Illustrations. Crown 8vo, 7s. 6d. cloth.

"The subject of ship electric lighting is one of vast importance in these days, and Mr. Urquhart is to be highly complimented for placing such a valuable work at the service of the practical
marine electrician."—The Steamship.

"Distinctly a book which of its kind stands almost alone, and for which there should be a
demand."—Electrical Review.

Country House Electric Lighting.

ELECTRIC LIGHT FOR COUNTRY HOUSES : A Practical
Handbook on the Erection and Running of Small Installations, with particulars of the Cost of Plant and Working. By J. H. KNIGHT. Crown 8vo,
1s. wrapper. [Just published.

Electric Lighting.

THE ELEMENTARY PRINCIPLES OF ELECTRIC LIGHT-
ING. By ALAN A. CAMPBELL SWINTON, Associate I.E.E. Third Edition,
Enlarged and Revised. With 16 Illustrations. Crown 8vo, 1s. 6d. cloth.

"Anyone who desires a short and thoroughly clear exposition of the elementary principles of
electric-lighting cannot do better than read this little work."—Bradford Observer.

Dynamic Electricity.

THE ELEMENTS OF DYNAMIC ELECTRICITY AND
MAGNETISM. By PHILIP ATKINSON, A.M., Ph.D., Author of "The Elements of Electric Lighting," &c. Cr. 8vo, with 120 Illustrations, 10s. 6d. cl.

Electric Motors, &c.

THE ELECTRIC TRANSFORMATION OF POWER and its
Application by the Electric Motor, including Electric Railway Construction.
By P. ATKINSON, A.M., Ph.D., Author of "The Elements of Electric Light-
ing," &c. With 96 Illustrations. Crown 8vo, 7s. 6d. cloth.

Dynamo Construction.

HOW TO MAKE A DYNAMO : A Practical Treatise for Amateurs.
Containing numerous Illustrations and Detailed Instructions for Construct-
ing a Small Dynamo, to Produce the Electric Light. By ALFRED CROFTS.
Fifth Edition, Revised and Enlarged. Crown 8vo, 2s. cloth.

"The instructions given in this unpretentious little book are sufficiently clear and explicit to
enable any amateur mechanic possessed of average skill and the usual tools to be found in an
amateur's workshop, to build a practical dynamo machine."—Electrician.

Text Book of Electricity.

THE STUDENT'S TEXT-BOOK OF ELECTRICITY. By
HENRY M. NOAD, F.R.S. 630 pages, with 470 Illustrations. Cheaper Edition.
Crown 8vo, 9s. cloth. [Just published.

Electricity.

A MANUAL OF ELECTRICITY : Including Galvanism, Mag-
netism, Dia-Magnetism, Electro-Dynamics. By HENRY M. NOAD, Ph D., F.R.S
Fourth Edition (1859). 8vo, £1 4s. cloth.

ARCHITECTURE, BUILDING, etc.

Building Construction.

PRACTICAL BUILDING CONSTRUCTION: A Handbook for Students Preparing for Examinations, and a Book of Reference for Persons Engaged in Building. By JOHN P. ALLEN, Surveyor, Lecturer on Building Construction at the Durham College of Science, Newcastle. Medium 8vo, 450 pages, with 1,000 Illustrations. 12s. 6d. cloth.   [*Just published.*

" This volume is one of the most complete expositions of building construction we have seen. It contains all that is necessary to prepare students for the various examinations in building construction."—*Building News.*

" The author depends nearly as much on his diagrams as on his type. The pages suggest the hand of a man of experience in building operations—and the volume must be a blessing to many teachers as well as to students. '—*The Architect.*

" The work is sure to prove a formidable rival to great and small competitors alike, and bids fair to take a permanent place as a favourite students' text-book. The large number of illustrations deserve particular mention for the great merit they possess for purposes of reference, in exactly corresponding to convenient scales."—*Jour. Inst. Brit. Archts.*

Masonry.

PRACTICAL MASONRY: A Guide to the Art of Stone Cutting. Comprising the Construction, Setting-Out, and Working of Stairs, Circular Work, Arches, Niches, Domes, Pendentives, Vaults, Tracery Windows, &c. For the Use of Students, Masons and other Workmen. By WILLIAM R. PURCHASE, Building Inspector to the Town of Hove. Royal 8vo, 134 pages, including 50 Lithographic Plates (about 400 separate Diagrams), 7s. 6d. cloth.   [*Just published.*

" The illustrations are well thought out and clear. The volume places within reach of the professional mason many useful data for solving the problems which present themselves day by day.' —*Glasgow Herald.*

The New Builder's Price Book, 1896.

LOCKWOOD'S BUILDER'S PRICE BOOK FOR 1896. A Comprehensive Handbook of the Latest Prices and Data for Builders, Architects, Engineers, and Contractors. By FRANCIS T. W. MILLER. 800 closely-printed pages, crown 8vo, 4s. cloth.

" This book is a very useful one, and should find a place in every English office connected with the building and engineering professions."—*Industries.*

" An excellent book of reference."—*Architect.*

" In its new and revised form this Price Book is what a work of this kind should be—comprehensive, reliable, well arranged, legible, and well bound."—*British Architect.*

New London Building Act, 1894.

THE LONDON BUILDING ACT, 1894; with the By-Laws and Regulations of the London County Council, and Introduction, Notes, Cases and Index. By ALEX. J. DAVID, B.A., LL.M. of the Inner Temple, Barrister-at-Law. Crown 8vo, 3s. 6d. cloth.   [*Just published.*

" To all architects and district surveyors and builders, Mr. David's manual will be welcome."— *Building News.*

" The volume will doubtless be eagerly consulted by the building fraternity."—*Illustrated Carpenter and Builder.*

Concrete.

CONCRETE: ITS NATURE AND USES. A Book for Architects, Builders, Contractors, and Clerks of Works. By GEORGE L. SUTCLIFFE, A.R.I.B.A. Crown 8vo, 7s. 6d. cloth.   [*Just published.*

" The author treats a difficult subject in a lucid manner. The manual fills a long-felt gap. It careful and exhaustive ; equally useful as a student's guide and a architect's book of reference." —*Journal of Royal Institution of British Architects.*

" There is room for this new book, which will probably be for some time the standard work on the subject for a builder's purpose."—*Glasgow Herald.*

" A thoroughly useful and comprehensive work."—*British Architect.*

Mechanics for Architects.

THE MECHANICS OF ARCHITECTURE: A Treatise on Applied Mechanics, especially Adapted to the Use of Architects. By E. W. TARN, M.A., Author of "The Science of Building," &c. Second Edition, Enlarged. Illust. with 125 Diagrams. Cr. 8vo, 7s. 6d. cloth. [*Just published.*

" The book is a very useful and helpful manual of architectural mechanics, and really contains sufficient to enable a careful and painstaking student to grasp the principles bearing upon the majority of building problems. . . . Mr. Tarn has added, by this volume, to the debt of gratitude which is owing to him by architectural students for the many valuable works which he has produced for their use."—*The Builder.*

" The mechanics in the volume are really mechanics, and are harmoniously wrought in with the distinctive professional manner proper to the subject. '—*The Schoolmaster.*   **H**

Designing Buildings.

THE DESIGN OF BUILDINGS: Being Elementary Notes on the Planning, Sanitation and Ornamentive Formation of Structures, based on Modern Practice. Illustrated with Nine Folding Plates. By W. WOODLEY, Assistant Master, Metropolitan Drawing Classes, &c. 8vo, 6s. cloth.

Sir Wm. Chambers's Treatise on Civil Architecture.

THE DECORATIVE PART OF CIVIL ARCHITECTURE. By Sir WILLIAM CHAMBERS, F.R.S. With Portrait, Illustrations, Notes, and an Examination of Grecian Architecture, by JOSEPH GWILT, F.S.A. Revised and Edited by W. H. LEEDS. 66 Plates, 4to, 21s. cloth.

Villa Architecture.

A HANDY BOOK OF VILLA ARCHITECTURE: Being a Series of Designs for Villa Residences in various Styles. With Outline Specifications and Estimates. By C. WICKES, Architect, Author of "The Spires and Towers of England," &c. 61 Plates, 4to, £1 11s. 6d. half-morocco.
" The whole of the designs bear evidence of their being the work of an artistic architect, and they will prove very valuable and suggestive."—*Building News.*

Text-Book for Architects.

THE ARCHITECT'S GUIDE: Being a Text-Book of Useful Information for Architects, Engineers, Surveyors, Contractors, Clerks of Works, &c. By F. ROGERS. Third Edition. Crown 8vo, 3s. 6d. cloth.
" As a text-book of useful information for architects, engineers, surveyors, &c., it would be hard to find a handier or more complete little volume."—*Standard.*

Linear Perspective.

ARCHITECTURAL PERSPECTIVE: The whole Course and Operations of the Draughtsman in Drawing a Large House in Linear Perspective. Illustrated by 43 Folding Plates. By F. O. FERGUSON. Second Edition, Enlarged. 8vo, 3s. 6d. boards.   [Just published.
" It is the most intelligible of the treatises on this ill treated subject that I have met with."—E. INGRESS BELL, Esq., in the *R.I.B.A. Journal.*

Architectural Drawing.

PRACTICAL RULES ON DRAWING, for the Operative Builder and Young Student in Architecture. By G. PYNE. 14 Plates, 4to, 7s. 6d., bds.

Vitruvius' Architecture.

THE ARCHITECTURE of MARCUS VITRUVIUS POLLIO. Translated by JOSEPH GWILT, F.S.A., F.R.A.S. New Edition, Revised by the Translator. With 23 Plates. Fcap. 8vo, 5s. cloth.

Designing, Measuring, and Valuing.

THE STUDENT'S GUIDE to the PRACTICE of MEASURING AND VALUING ARTIFICERS' WORK. Containing Directions for taking Dimensions, Abstracting the same, and bringing the Quantities into Bill, with Tables of Constants for Valuation of Labour, and for the Calculation of Areas and Solidities. Originally edited by EDWARD DOBSON, Architect. With Additions by E. WYNDHAM TARN, M.A. Sixth Edition. With 8 Plates and 63 Woodcuts. Crown 8vo, 7s. 6d. cloth.
" This edition will be found the most complete treatise on the principles of measuring and valuing artificers' work that has yet been published."—*Building News.*

Pocket Estimator and Technical Guide.

THE POCKET TECHNICAL GUIDE, MEASURER, AND ESTIMATOR FOR BUILDERS AND SURVEYORS. Containing Technical Directions for Measuring Work in all the Building Trades, Complete Specifications for Houses, Roads, and Drains, and an easy Method of Estimating the parts of a Building collectively. By A. C. BEATON. Seventh Edit. Waistcoat-pocket size, 1s. 6d. leather, gilt edges.
" No builder, architect, surveyor, or valuer should be without his ' Beaton.' "—*Building News.*

Donaldson on Specifications.

THE HANDBOOK OF SPECIFICATIONS; or, Practical Guide to the Architect, Engineer, Surveyor, and Builder, in drawing up Specifications and Contracts for Works and Constructions. Illustrated by Precedents of Buildings actually executed by eminent Architects and Engineers. By Professor T. L. DONALDSON, P.R.I.B.A., &c. New Edition. 8vo. with upwards of 1,000 pages of Text, and 33 Plates. £1 11s. 6d. cloth.
" Valuable as a record, and more valuable still as a book of precedents. . . . Suffice it to say that Donaldson's ' Handbook of Specifications ' must be bought by all architects."—*Builder.*

Bartholomew and Rogers' Specifications.

SPECIFICATIONS FOR PRACTICAL ARCHITECTURE.
A Guide to the Architect, Engineer, Surveyor, and Builder. With an Essay on the Structure and Science of Modern Buildings. Upon the Basis of the Work by ALFRED BARTHOLOMEW, thoroughly Revised, Corrected, and greatly added to by FREDERICK ROGERS, Architect. Third Edition, Revised, with Additions. With numerous Illustrations. Medium 8vo, 15s. cloth.

" The collection of specifications prepared by Mr. Rogers on the basis of Bartholomew's work is too well known to need any recommendation from us. It is one of the books with which every young architect must be equipped."—*Architect.*

House Building and Repairing.

· *THE HOUSE-OWNER'S ESTIMATOR* ; or, What will it Cost to Build, Alter, or Repair? A Price Book for Unprofessional People, as well as the Architectural Surveyor and Builder. By J. D. SIMON. Edited by F. T. W. MILLER, A.R.I.B.A. Fourth Edition. Crown 8vo, 3s. 6d. cloth.

"In two years it will repay its cost a hundred times over."—*Field.*

Construction.

THE SCIENCE OF BUILDING : An Elementary Treatise on the Principles of Construction. By E. WYNDHAM TARN, M.A., Architect. Third Edition, Revised and Enlarged. With 59 Engravings. Fcap. 8vo, 4s. cl.

A very valuable book, which we strongly recommend to all students."—*Builder.*

Building ; Civil and Ecclesiastical.

A BOOK ON BUILDING, Civil and Ecclesiastical, including Church Restoration ; with the Theory of Domes and the Great Pyramid, &c. By Sir EDMUND BECKETT, Bart., LL.D., F.R.A.S. Fcap. 8vo, 5s. cloth.

" A book which is always amusing and nearly always instructive."—*Times.*

House Building.

DWELLING HOUSES, THE ERECTION OF. Illustrated by a Perspective View, Plans, Elevations and Sections of a Pair of Semi-Detached Villas, with the Specification, Quantities and Estimates. By S. H. BROOKS, Architect. Seventh Edition, thoroughly Revised. 12mo, 2s. 6d. cloth. [*Just published*

Sanitary Houses, etc.

THE SANITARY ARRANGEMENT OF DWELLING-HOUSES: A Handbook for Householders and Owners of Houses. By A. J. WALLIS-TAYLER, A.M. Inst. C.E. With numerous Illustrations. Crown 8vo, 2s. 6d. cloth. [*Just published.*

" This book will be largely read ; it will be of considerable service to the public. It is well arranged, easily read, and for the most part devoid of technical terms."—*Lancet.*

Ventilation of Buildings.

VENTILATION. A Text Book to the Practice of the Art of Ventilating Buildings. By W. P. BUCHAN, R.P. 12mo, 4s. cloth.

" Contains a great amount of useful practical information, as thoroughly interesting as it is technically reliable."—*British Architect.*

The Art of Plumbing.

PLUMBING. A Text Book to the Practice of the Art or Craft of the Plumber. By WILLIAM PATON BUCHAN, R.P. Sixth Edition. 4s. cloth.

" A text-book which may be safely put in the hands of every young plumber."—*Builder.*

Geometry for the Architect, Engineer, etc.

PRACTICAL GEOMETRY, for the Architect, Engineer, and Mechanic. Giving Rules for the Delineation and Application of various Geometrical Lines, Figures and Curves. By E. W. TARN, M.A., Architect. 8vo, 9s. cloth.

" No book with the same objects in view has ever been published in which the clearness of the rules laid down and the illustrative diagrams have been so satisfactory."—*Scotsman.*

The Science of Geometry.

THE GEOMETRY OF COMPASSES; or, Problems Resolved by the mere Description of Circles, and the use of Coloured Diagrams and Symbols. By OLIVER BYRNE. Coloured Plates. Crown 8vo, 3s. 6d. cloth.

CARPENTRY, TIMBER, etc.

Tredgold's Carpentry, Revised & Enlarged by Tarn.

THE ELEMENTARY PRINCIPLES OF CARPENTRY.
A Treatise on the Pressure and Equilibrium of Timber Framing, the Resistance of Timber, and the Construction of Floors, Arches, Bridges, Roofs, Uniting Iron and Stone with Timber, &c. To which is added an Essay on the Nature and Properties of Timber, &c., with Descriptions of the kinds of Wood used in Building; also numerous Tables of the Scantlings of Timber for different purposes, the Specific Gravities of Materials, &c. By THOMAS TREDGOLD, C.E. With an Appendix of Specimens of Various Roofs of Iron and Stone, Illustrated. Seventh Edition, thoroughly revised and considerably enlarged by E. WYNDHAM TARN, M.A., Author of "The Science of Building," &c. With 61 Plates, Portrait of the Author, and several Woodcuts. In One large Vol., 4to, price £1 5s. cloth.
"Ought to be in every architect's and every builder's library."—*Builder.*
"A work whose monumental excellence must commend it wherever skilful carpentry is concerned. The author's principles are rather confirmed than impaired by time. The additional plates are of great intrinsic value."—*Building News.*

Carpentry.

CARPENTRY AND JOINERY. The Elementary Principles of Carpentry. Chiefly composed from the Standard Work of THOMAS TREDGOLD, C.E. With Additions, and a TREATISE ON JOINERY by E. W. TARN, M.A. Fifth Edition, Revised and Extended. 12mo, 3s. 6d. cloth.
** ATLAS of Thirty-five Plates to accompany and illustrate the foregoing book. With Descriptive Letterpress. 4to, 6s. cloth.
"These two volumes form a complete treasury of carpentry and joinery, and should be in the hands of every carpenter and joiner in the empire."—*Iron.*

Woodworking Machinery.

WOODWORKING MACHINERY: Its Rise, Progress, and Construction. With Hints on the Management of Saw Mills and the Economical Conversion of Timber. Illustrated with Examples of Recent Designs by leading English, French, and American Engineers. By M. POWIS BALE, A.M.Inst.C.E., M.I.M.E. Second Edition, Revised, with large Additions. Large crown 8vo, 440 pp., 9s. cloth. [*Just published.*
"Mr. Bale is evidently an expert on the subject and he has collected so much information that the book is all-sufficient for builders and others engaged in the conversion of timber."—*Architect.*
"The most comprehensive compendium of wood-working machinery we have seen. The author is a thorough master of his subject."—*Building News.*

Saw Mills.

SAW MILLS: Their Arrangement and Management, and the Economical Conversion of Timber. (A Companion Volume to "Woodworking Machinery.") By M. POWIS BALE. Crown 8vo, 10s. 6d. cloth.
"The *administration* of a large sawing establishment is discussed and the subject examined from a financial standpoint. Hence the size, shape, order, and disposition of saw-mills and the like are gone into in detail, and the course of the timber is traced from its reception to its delivery in its converted state. We could not desire a more complete or practical treatise."—*Builder.*

Nicholson's Carpentry.

THE CARPENTER'S NEW GUIDE; or, Book of Lines for Carpenters; comprising all the Elementary Principles essential for acquiring a knowledge of Carpentry. Founded on the late PETER NICHOLSON'S Standard Work. New Edition, Revised by A. ASHPITEL, F.S.A. With Practical Rules on Drawing, by G. PYNE. With 74 Plates, 4to, £1 1s. cloth.

Circular Work.

CIRCULAR WORK IN CARPENTRY AND JOINERY: A Practical Treatise on Circular Work of Single and Double Curvature. By GEORGE COLLINGS. With Diagrams. Second Edit. 12mo, 2s. 6d. cloth limp.
"An excellent example of what a book of this kind should be. Cheap in price, clear in definition and practical in the examples selected."—*Builder.*

Handrailing.

HANDRAILING COMPLETE IN EIGHT LESSONS. On the Square-Cut System. By J. S. GOLDTHORP, Teacher of Geometry and Building Construction at the Halifax Mechanic's Institute. With Eight Plates and over 150 Practical Exercises. 4to, 3s. 6d. cloth.
"Likely to be of considerable value to joiners and others who take a pride in good work. We heartily commend it to teachers and students."—*Timber Trades Journal.*

Handrailing and Stairbuilding.

A PRACTICAL TREATISE ON HANDRAILING : Showing New and Simple Methods for Finding the Pitch of the Plank, Drawing the Moulds, Bevelling, Jointing-up, and Squaring the Wreath. By GEORGE COLLINGS. Second Edition, Revised and Enlarged, to which is added A TREATISE ON STAIRBUILDING. 12mo, 2s. 6d. cloth limp.

"Will be found of practical utility in the execution of this difficult branch of joinery."—*Builder.*
" Almost every difficult phase of this somewhat intricate branch of joinery is elucidated by the aid of plates and explanatory letterpress."—*Furniture Gazette.*

Timber Merchant's Companion.

THE TIMBER MERCHANT'S AND BUILDER'S COMPANION. Containing New and Copious Tables of the Reduced Weight and Measurement of Deals and Battens, of all sizes, from One to a Thousand Pieces, and the relative Price that each size bears per Lineal Foot to any given Price per Petersburg Standard Hundred; the Price per Cube Foot of Square Timber to any given Price per Load of 50 Feet; the proportionate Value of Deals and Battens by the Standard, to Square Timber by the Load of 50 Feet; the readiest mode of ascertaining the Price cf Scantling per Lineal Foot of any size, to any given Figure per Cube Foot, &c. &c. By WILLIAM DOWSING. Fourth Edition, Revised and Corrected. Cr. 8vo, 3s. cl.

"Everything is as concise and clear as it can possibly be made. There can be no doubt that every timber merchant and builder ought to possess it."—*Hull Advertiser.*
" We are glad to see a fourth edition of these admirable tables, which for correctness and simplicity of arrangement leave nothing to be desired."—*Timber Trades Journal.*

Practical Timber Merchant.

THE PRACTICAL TIMBER MERCHANT. Being a Guide for the use of Building Contractors, Surveyors, Builders, &c., comprising useful Tables for all purposes connected with the Timber Trade, Marks of Wood, Essay on the Strength of Timber, Remarks on the Growth of Timber, &c. By W. RICHARDSON. Second Edi ion. Fcap. 8vo, 3s. 6d. cloth.

"This handy manual contains much valuable information for the use of timber merchants, builders, foresters, and all others connected with the growth, sale, and manufacture of timber."—*Journal of Forestry.*

Packing-Case Makers, Tables for.

PACKING-CASE TABLES ; showing the number of Superficial Feet in Boxes or Packing-Cases, from six inches square and upwards. By W. RICHARDSON, Timber Broker. Third Edition. Oblong 4to, 3s. 6d. cl.

"Invaluable labour-saving tables."—*Ironmonger.*
"Will save much labour and calculation."—*Grocer.*

Superficial Measurement.

THE TRADESMAN'S GUIDE TO SUPERFICIAL MEASUREMENT. Tables calculated from 1 to 200 inches in length, by 1 to 108 inches in breadth. For the use of Architects, Surveyors, Engineers, Timber Merchants, Builders, &c. By JAMES HAWKINGS. Fourth Edition. Fcap., 3s. 6d. cloth.

"A useful collection of tables to facilitate rapid calculation of surfaces. The exact area of any surface of which the limits have been ascertained can be instantly determined. The book will be found of the greatest utility to all engaged in building operations."—*Scotsman.*
" These tables will be found of great assistance to all who require to make calculations in superficial measurement."—*English Mechanic.*

Forestry.

THE ELEMENTS OF FORESTRY. Designed to afford Information concerning the Planting and Care of Forest Trees for Ornament or Profit, with Suggestions upon the Creation and Care of Woodlands. By F. B. HOUGH. Large crown 8vo, 10s. cloth.

Timber Importer's Guide.

THE TIMBER IMPORTER'S, TIMBER MERCHANT'S, AND BUILDER'S STANDARD GUIDE. By RICHARD E. GRANDY. Comprising an Analysis of Deal Standards, Home and Foreign, with Comparative Values and Tabular Arrangements for fixing Net Landed Cost on Baltic and North American Deals, including all intermediate Expenses, Freight, Insurance, &c. &c. Together with copious Information for the Retailer and Builder. Third Edition, Revised. 12mo, 2s. cloth limp.

"Everything it pretends to be : built up gradually, it leads one from a forest to a treenail, and throws in, as a makeweight, a host of material concerning bricks, columns, cisterns, &c."—*English Mechanic.*

DECORATIVE ARTS, etc.

Woods and Marbles (Imitation of).

SCHOOL OF PAINTING FOR THE IMITATION OF WOODS AND MARBLES, as Taught and Practised by A. R. VAN DER BURG and P. VAN DER BURG, Directors of the Rotterdam Painting Institution. Royal folio, 18¼ by 12¼ in., Illustrated with 24 full-size Coloured Plates; also 12 plain Plates, comprising 154 Figures. Second and Cheaper Edition. Price £1 11s. 6d.

List of Plates.

1. Various Tools required for Wood Painting—2, 3. Walnut: Preliminary Stages of Graining and Finished Specimen—4. Tools used for Marble Painting and Method of Manipulation—6. St. Remi Marble: Earlier Operations and Finished Specimen—7. Methods of Sketching different Grains, Knots, &c.—8, 9. Ash: Preliminary Stages and Finished Specimen—10. Methods of Sketching Marble Grains—11, 12. Breche Marble: Preliminary Stages of Working and Finished Specimen—13. Maple: Methods of Producing the different Grains—14, 15. Bird's-eye Maple: Preliminary Stages and Finished Specimen—16. Methods of Sketching the different Species of White Marble—17, 18. White Marble: Preliminary Stages of Process and Finished Specimen—19. Mahogany: Specimens of various Grains and Methods of Manipulation—20, 21. Mahogany: Earlier Stages and Finished Specimen—22, 23, 24. Sienna Marble: Varieties of Grain, Preliminary Stages and Finished Specimen—25, 26, 27. Juniper Wood: Methods of producing Grain, &c.: Preliminary Stages and Finished Specimen—28, 29, 30. Vert de Mer Marble: Varieties of Grain and Methods of Working Unfinished and Finished Specimens—31. 32. 33. Oak: Varieties of Grain, Tools Employed, and Methods of Manipulation, Preliminary Stages and Finished Specimen—34, 35, 36. Waulsort Marble: Varieties of Grain, Unfinished and Finished Specimens.

"Those who desire to attain skill in the art of painting woods and marbles will find advantage in consulting this book. . . . Some of the Working Men's Clubs should give their young men the opportunity to study it."—*Builder.*

"A comprehensive guide to the art. The explanations of the processes, the manipulation and management of the colours, and the beautifully executed plates will not be the least valuable to the student who aims at making his work a faithful transcript of nature."—*Building News.*

House Decoration.

ELEMENTARY DECORATION. A Guide to the Simpler Forms of Everyday Art. Together with *PRACTICAL HOUSE DECORATION.* By JAMES W. FACEY. With numerous Illustrations. In One Vol., 5s. strongly half bound

House Painting, Graining, etc.

HOUSE PAINTING, GRAINING, MARBLING, AND SIGN WRITING, A Practical Manual of. By ELLIS A. DAVIDSON. Sixth Edition With Coloured Plates and Wood Engravings. 12mo, 6s. cloth boards.

"A mass of information, of use to the amateur and of value to the practical man."—*English Mechanic.*

Decorators, Receipts for.

THE DECORATOR'S ASSISTANT : A Modern Guide to Decorative Artists and Amateurs, Painters, Writers, Gilders, &c. Containing upwards of 600 Receipts, Rules and Instructions ; with a variety of Information for General Work connected with every Class of Interior and Exterior Decorations, &c. Sixth Edition. 152 pp., crown 8vo, 1s. in wrapper

"Full of receipts of value to decorators, painters, gilders, &c. The book contains the gist of larger treatises on colour and technical processes. It would be difficult to meet with a work so full of varied information on the painter's art."—*Building News.*

Moyr Smith on Interior Decoration.

ORNAMENTAL INTERIORS, ANCIENT AND MODERN. By J. MOYR SMITH. Super-royal 8vo, with 32 full-page Plates and numerous smaller Illustrations, handsomely bound in cloth, gilt top, price 18s.

"The book is well illustrated and handsomely got up, and contains some true criticism and a good many good examples of decorative treatment."—*The Builder*

British and Foreign Marbles.

MARBLE DECORATION and the Terminology of British and Foreign Marbles. A Handbook for Students. By GEORGE H. BLAGROVE, Author of "Shoring and its Application," &c. With 28 Illustrations. Crown 8vo, 3s. 6d. cloth.

"This most useful and much wanted handbook should be in the hands of every architect and builder."—*Building World.*

"A carefully and usefully written treatise; the work is essentially practical."—*Scotsman.*

Marble Working, etc.

MARBLE AND MARBLE WORKERS: A Handbook for Architects, Artists, Masons, and Students. By ARTHUR LEE, Author of "A Visit to Carrara," "The Working of Marble," &c. Small crown 8vo, 2s. cloth.

"A really valuable addition to the technical literature of architects and masons."—*Building News.*

DELAMOTTE'S WORKS ON ILLUMINATION AND ALPHABETS.

A PRIMER OF THE ART OF ILLUMINATION, for the Use of Beginners: with a Rudimentary Treatise on the Art, Practical Directions for its Exercise, and Examples taken from Illuminated MSS., printed in Gold and Colours. By F. DELAMOTTE. New and Cheaper Edition. Small 4to, 6s. ornamental boards.

"The examples of ancient MSS. recommended to the student, which, with much good sense, the author chooses from collections accessible to all, are selected with judgment and knowledge, as well as taste."—*Athenæum.*

ORNAMENTAL ALPHABETS, Ancient and Mediæval, from the Eighth Century, with Numerals; including Gothic, Church-Text, large and small, German, Italian, Arabesque, Initials for Illumination, Monograms, Crosses, &c. &c., for the use of Architectural and Engineering Draughtsmen Missal Painters, Masons, Decorative Painters, Lithographers, Engravers Carvers, &c. &c. Collected and Engraved by F. DELAMOTTE, and printed in Colours. New and Cheaper Edition. Royal 8vo, oblong, 2s. 6d. ornamental boards.

"For those who insert enamelled sentences round gilded chalices, who blazon shop legends over shop-doors, who letter church walls with pithy sentences from the Decalogue, this book will be useful."—*Athenæum.*

EXAMPLES OF MODERN ALPHABETS, Plain and Ornamental; including German, Old English, Saxon, Italic, Perspective, Greek, Hebrew, Court Hand, Engrossing, Tuscan, Riband, Gothic, Rustic, and Arabesque; with several Original Designs, and an Analysis of the Roman and Old English Alphabets, large and small, and Numerals, for the use of Draughtsmen, Surveyors, Masons, Decorative Painters, Lithographers, Engravers, Carvers, &c. Collected and Engraved by F. DELAMOTTE, and printed in Colours. New and Cheaper Edition. Royal 8vo, oblong, 2s. 6d. ornamental boards.

"There is comprised in it every possible shape into which the letters of the alphabet and numerals can be formed, and the talent which has been expended in the conception of the various plain and ornamental letters is wonderful."—*Standard.*

MEDIÆVAL ALPHABETS AND INITIALS FOR ILLUMI-NATORS. By F. G. DELAMOTTE. Containing 21 Plates and Illuminated Title, printed in Gold and Colours. With an Introduction by J. WILLIS BROOKS. Fourth and Cheaper Edition. Small 4to, 4s. ornamental boards.

"A volume in which the letters of the alphabet come forth glorified in gilding and all the colours of the prism interwoven and intermingled."—*Sun.*

THE EMBROIDERER'S BOOK OF DESIGN. Containing Initials, Emblems, Cyphers, Monograms, Ornamental Borders, Ecclesiastical Devices, Mediæval and Modern Alphabets, and National Emblems. Collected by F. DELAMOTTE, and printed in Colours. Oblong royal 8vo, 1s. 6d. ornamental wrapper.

"The book will be of great assistance to ladies and young children who are endowed with the art of plying the needle in this most ornamental and useful pretty work."—*East Anglian Times.*

Wood Carving.

INSTRUCTIONS IN WOOD-CARVING, for Amateurs: with Hints on Design. By A LADY. With Ten Plates. New and Cheaper Edition. Crown 8vo, 2s. in emblematic wrapper.

"The handicraft of the wood-carver, so well as a book can impart it, may be learnt from 'A Lady's' publication."—*Athenæum.*

NATURAL SCIENCE, etc.

The Heavens and their Origin.

THE VISIBLE UNIVERSE: Chapters on the Origin and Construction of the Heavens. By J. E. GORE, F.R.A.S. Illustrated by 6 Stellar Photographs and 12 Plates. 8vo, 16s. cloth.

"A valuable and lucid summary of recent astronomical theory, rendered more valuable and attractive by a series of stellar photographs and other illustrations."—*The Times.*

"In presenting a clear and concise account of the present state of our knowledge, Mr. Gore has made a valuable addition to the literature of the subject."—*Nature.*

"As interesting as a novel, and instructive withal; the text being made still more luminous by stellar photographs and other illustrations. . . . A most valuable book."—*Manchester Examiner.*

"One of the finest works on astronomical science that has recently appeared in our language."
Leeds Mercury

The Constellations.

STAR GROUPS: A Student's Guide to the Constellations. By J. ELLARD GORE, F.R.A.S., M.R.I.A., &c., Author of "The Visible Universe," "The Scenery of the Heavens." With 30 Maps. Small 4to, 5s. cloth, silvered.

"A knowledge of the principal constellations visible in our latitudes may be easily acquired from the thirty maps and accompanying text contained in this work."—*Nature.*

"The volume contains thirty maps showing stars of the sixth magnitude—the usual naked-eye limit—and each is accompanied by a brief commentary, adapted to facilitate recognition and bring to notice objects of special interest. For the purpose of a preliminary survey of the 'midnight pomp' of the heavens, nothing could be better than a set of delineations averaging scarcely twenty square inches in area, and including nothing that cannot at once be identified."—*Saturday Review.*

"A very compact and handy guide to the constellations."—*Athenæum.*

Astronomical Terms.

AN ASTRONOMICAL GLOSSARY; or, Dictionary of Terms used in Astronomy. With Tables of Data and Lists of Remarkable and Interesting Celestial Objects. By J. ELLARD GORE, F.R.A.S., Author of "The Visible Universe," &c. Small crown 8vo, 2s. 6d. cloth.

"A very useful little work for beginners in astronomy, and not to be despised by more advanced students."—*The Times.*

"Astronomers of all kinds will be glad to have it for reference."—*Guardian.*

The Microscope.

THE MICROSCOPE: Its Construction and Management, including Technique, Photo-micrography, and the Past and Future of the Microscope. By Dr. HENRI VAN HEURCK. Re-Edited and Augmented from the Fourth French Edition, and Translated by WYNNE E. BAXTER, F.G.S. 400 pages, with upwards of 250 Woodcuts. Imp. 8vo, 18s. cloth.

"A translation of a well-known work, at once popular and comprehensive."—*Times.*

"The translation is as felicitious as it is accurate."—*Nature.*

The Microscope.

PHOTO-MICROGRAPHY. By Dr. H. VAN HEURCK. Extracted from the above Work. Royal 8vo, with Illustrations, 1s. sewed.

Astronomy.

ASTRONOMY. By the late Rev. ROBERT MAIN, F.R.S. Third Edition, Revised, by WM. T. LYNN, B.A., F.R.A.S., 12mo, 2s. cloth.

"A sound and simple treatise, and a capital book for beginners."—*Knowledge.*

Recent and Fossil Shells.

A MANUAL OF THE MOLLUSCA: Being a Treatise on Recent and Fossil Shells. By S. P. WOODWARD, A.L.S., F.G.S. With an Appendix on Recent and Fossil Conchological Discoveries, by RALPH TATE, A.L.S., F.G.S. With 23 Plates and upwards of 300 Woodcuts. Reprint of Fourth Edition, 1880. Crown 8vo, 7s. 6d. cloth

"A most valuable storehouse of conchological and geological information."—*Science Gossip.*

Geology and Genesis.

THE TWIN RECORDS OF CREATION; or, Geology and Genesis: their Perfect Harmony and Wonderful Concord. By GEORGE W. VICTOR LE VAUX. Fcap. 8vo, 5s. cloth.

"A valuable contribution to the evidences of Revelation, and disposes very conclusively of the arguments of those who would set God's Works against God's Word. No real difficulty is shirked, and no sophistry is left unexposed."—*The Rock.*

Geology.

RUDIMENTARY TREATISE ON GEOLOGY, PHYSICAL AND HISTORICAL. With especial reference to the British series of Rocks. By R. TATE, F.G.S. With 250 Illustrations. 12mo, 5s. cloth boards.

DR. LARDNER'S COURSE OF NATURAL PHILOSOPHY.

HANDBOOK OF MECHANICS. Re-written and Enlarged by B. LOEWY, F.R.A.S. Post 8vo, 6s. cloth.

"Mr. Loewy has carefully revised the book, and brought it up to modern requirements."— *Nature.*

HANDBOOK OF HYDROSTATICS & PNEUMATICS. Enlarged by R. LOEWY, F.R.A.S. Post 8vo, 5s. cloth.

"For those 'who desire to attain an accurate knowledge of physical science without the profound methods of mathematical investigation,' this work is well adapted."—*Chemical News.*

HANDBOOK OF HEAT. Edited and almost entirely Re-written by BENJAMIN LOEWY, F.R.A.S., &c. Post 8vo, 6s. cloth.

"The style is always clear and precise, and conveys instruction without leaving any cloudiness or lurking doubts behind."—*Engineering.*

HANDBOOK OF OPTICS. By Dr. LARDNER. Edited by T. O. HARDING, B.A. Post 8vo, 5s. cloth.

"Written by an able scientific writer and beautifully illustrated."—*Mechanic's Magazine.*

HANDBOOK OF ELECTRICITY AND MAGNETISM. By Dr. LARDNER. Edited by G. C. FOSTER, B.A. Post 8vo, 5s. cloth.

"The book could not have been entrusted to anyone better calculated to preserve the terse and lucid style of Lardner."—*Popular Science Review.*

HANDBOOK OF ASTRONOMY. By Dr. LARDNER. Fourth Edition by E. DUNKIN, F.R.A.S. Post 8vo, 9s. 6d. cloth.

"Probably no other book contains the same amount of information in so compendious and well-arranged a form—certainly none at the price at which this is offered to the public."—*Athenæum.*

"We can do no other than pronounce this work a most valuable manual of astronomy, and we strongly recommend it to all who wish to acquire a general—but at the same time correct—acquaintance with this sublime science."—*Quarterly Journal of Science.*

DR. LARDNER'S MUSEUM OF SCIENCE AND ART,

THE MUSEUM OF SCIENCE AND ART. Edited by Dr. LARDNER. With upwards of 1,200 Engravings on Wood. In 6 Double Volumes, £1 1s. in a new and elegant cloth binding; or handsomely bound in half-morocco, 31s. 6d.

"A cheap and interesting publication, alike informing and attractive. The papers combine subjects of importance and great scientific knowledge, considerable inductive powers, and a popular style of treatment."—*Spectator.*

The 'Museum of Science and Art' is the most valuable contribution that has ever been made to the Scientific Instruction of every class of society."—Sir DAVID BREWSTER, in the *North British Review.*

*** *Separate books formed from the above, fully Illustrated, suitable for Workmen's Libraries, Science Classes, etc.*

Common Things Explained. 5s.	***Steam and its Uses.*** 2s. cloth.
The Microscope. 2s. cloth.	***Popular Astronomy.*** 4s. 6d. cloth.
Popular Geology. 2s. 6d. cloth.	***The Bee and White Ants.*** 2s. cloth.
Popular Physics. 2s. 6d. cloth.	***The Electric Telegraph.*** 1s.

Dr. Lardner's School Handbooks.

NATURAL PHILOSOPHY FOR SCHOOLS. Fcap. 8vo, 3s. 6d

"A very convenient class-book for junior students in private schools."—*British Quarterly Review.*

ANIMAL PHYSIOLOGY FOR SCHOOLS. Fcap. 8vo, 3s. 6d.

"Clearly written, well arranged, and excellently illustrated."—*Gardener's Chronicle.*

THE ELECTRIC TELEGRAPH. By Dr. LARDNER. Revised by E. B. BRIGHT, F.R.A.S. Fcap. 8vo, 2s. 6d. cloth.

"One of the most readable books extant on the Electric Telegraph."—*English Mechanic.*

D

CHEMICAL MANUFACTURES, CHEMISTRY.

Refrigerating, etc.

REFRIGERATING AND ICE-MAKING MACHINERY:
A Descriptive Treatise for the Use of Persons Employing Refrigerating
and Ice-Making Installations, and others. By A. J. WALLIS-TAYLER, C.E.
Assoc. Member Inst. C.E. With Illustrations. Crown 8vo, 7s. 6d. cloth.
[*Just published.*

"Practical, explicit and profusely illustrated."—*Glasgow Herald.*

Chemistry for Engineers, etc.

ENGINEERING CHEMISTRY: A Practical Treatise for the
Use of Analytical Chemists, Engineers, Iron Masters, Iron Founders,
Students, and others. Comprising Methods of Analysis and Valuation of the
Principal Materials used in Engineering Work, with numerous Analyses,
Examples, and Suggestions. By H. JOSHUA PHILLIPS, F.I.C., F.C.S.
formerly Analytical and Consulting Chemist to the Great Eastern Railway.
Second Edition, Revised and Enlarged. Crown 8vo, 400 pp., with Illustra-
tions, 10s. 6d. cloth. [*Just published.*

"In this work the author has rendered no small service to a numerous body of practical men.
. . . The analytical methods may be pronounced most satisfactory, being as accurate as the
despatch required of engineering chemists permits."—*Chemical News.*
"The book will be very useful to those who require a handy and concise *resume* of approved
methods of analysing and valuing metals, oils, fuels, &c. It is, in fact, a work for chemists, a guide
to the routine of the engineering laboratory. . . . The book is full of good things. As a handbook of technical analysis, it is very welcome."—*Builder.*
"The analytical methods given are, as a whole, such as are likely to give rapid and trustworthy results in experienced hands. There is much excellent descriptive matter in the work, the
chapter on ' Oils and Lubrication ' being specially noticeable in this respect."—*Engineer.*

Manufacture of Explosives.

NITRO-EXPLOSIVES: A Practical Treatise concerning the
Properties, Manufacture, and Analysis of Nitrated Substances; including
the Fulminates, Smokeless Powders, and Celluloid. By P. GERALD SANFORD,
F.I.C., F.C.S. With numerous Illustrations. Crown 8vo, 9s. cloth.
[*Just published.*

"Mr. Sanford goes steadily through the whole list of explosives commonly used. He names
any given explosive, and tells us of what it is composed and how it is manufactured. The book is
excellent throughout, and we heartily recommend it."—*The Engineer.*

Explosives and Dangerous Goods.

DANGEROUS GOODS: Their Sources and Properties, Modes
of Storage, and Transport. With Notes and Comments on Accidents arising therefrom, together with the Government and Railway Classifications,
Acts of Parliament, &c. A Guide for the use of Government and Railway
Officials, Steamship Owners, Insurance Companies and Manufacturers and
users of Explosives and Dangerous Goods. By H. JOSHUA PHILLIPS, F.I.C.,
F.C.S., Author of "Engineering Chemistry, &c." Crown 8vo, 350 pp., 9s.
cloth. [*Just published.*

Explosives.

A HANDBOOK ON MODERN EXPLOSIVES. Being a
Practical Treatise on the Manufacture and Application of Dynamite, Gun-
Cotton, Nitro-Glycerine, and other Explosive Compounds. Including the
Manufacture of Collodion-Cotton. By M. EISSLER, Author of " The Metal-
lurgy of Gold," &c. Crown 8vo, 10s. 6d. cloth.
"Useful not only to the miner, but also to officers of both services to whom blasting and the
use of explosives generally may at any time become a necessary auxiliary."—*Nature.*
"A veritable mine of information on the subject of explosives employed for military, mining
and blasting purposes."—*Army and Navy Gazette.*

Alkali Trade, Manufacture of Sulphuric Acid, etc.

A MANUAL OF THE ALKALI TRADE, including the
Manufacture of Sulphuric Acid, Sulphate of Soda, and Bleaching Powder.
By JOHN LOMAS, Alkali Manufacturer, Newcastle-upon-Tyne and London.
With 232 Illustrations and Working Drawings, and containing 390 pages of
Text. Second Edition, with Additions. Super-royal 8vo, £1 10s. cloth.
"This book is written by a manufacturer for manufacturers. The working details of the most
approved forms of apparatus are given, and these are accompanied by no less than 232 wood en
gravings, all of which may be used for the purposes of construction. Every step in the manu-
facture is very fully described in this manual, and each improvement explained."—*Athenæum.*

The Blowpipe.

THE BLOWPIPE IN CHEMISTRY, MINERALOGY, AND GEOLOGY. Containing all known Methods of Anhydrous Analysis, many Working Examples, and Instructions for Making Apparatus. By Lieut.-Colonel W. A. Ross, R.A., F.G.S. With 120 Illustrations. Second Edition, Revised and Enlarged. Crown 8vo, 5s. cloth.

"The student who goes through the course of experimentation here laid down will gain a better insight into inorganic chemistry and mineralogy than if he had 'got up' any of the best text-books, and passed any number of examinations in their contents."—*Chemical News.*

Commercial Chemical Analysis.

THE COMMERCIAL HANDBOOK OF CHEMICAL ANALYSIS; or, Practical Instructions for the determination of the Intrinsic or Commercial Value of Substances used in Manufactures, in Trades, and in the Arts. By A. NORMANDY. New Edition, by H.M. NOAD, F.R.S. Crown 8vo, 12s. 6d. cloth.

"We strongly recommend this book to our readers as a guide, alike indispensable to the housewife as to the pharmaceutical practitioner."—*Medical Times.*

Dye-Wares and Colours.

THE MANUAL OF COLOURS AND DYE-WARES : Their Properties, Applications, Valuations, Impurities, and Sophistications. For the use of Dyers, Printers, Drysalters, Brokers, &c. By J. W. SLATER. Second Edition, Revised and greatly Enlarged. Crown 8vo, 7s. 6d. cloth.

"A complete encyclopædia of the *materia tinctoria.* The information given respecting each article is full and precise, and the methods of determining their value are given with clearness, an 1 are practical as well as valuable."—*Chemist and Druggist.*

Modern Brewing and Malting.

A HANDYBOOK FOR BREWERS : Being a Practical Guide to the Art of Brewing and Malting. Embracing the Conclusions of Modern Research which bear upon the Practice of Brewing. By HERBERT EDWARDS WRIGHT, M.A. Crown 8vo, 530 pp., 12s. 6d. cloth.

"May be consulted with advantage by the student who is preparing himself for examinational tests, while the scientific brewer will find in it a *resume* of all the most important discoveries of modern times. The work is written throughout in a clear and concise manner, and the author takes great care to discriminate between vague theories and well-established facts."—*Brewers' Journal*

"We have great pleasure in recommending this handybook, and have no hesitation in saying that it is one of the best—if not the best—which has yet been written on the subject of beer-brewing in this country, and it should have a place on the shelves of every brewer's library." —*The Brewer's Guardian.*

Analysis and Valuation of Fuels.

FUELS: SOLID, LIQUID, AND GASEOUS, Their Analysis and Valuation. For the Use of Chemists and Engineers. By H. J. PHILLIPS, F.C.S., formerly Analytical and Consulting Chemist to the Great Eastern Railway. Third Edition. Small crown 8vo, 2s. cloth.

"Ought to have its place in the laboratory of every metallurgical establishment, and wherever fuel is used on a large scale."—*Chemical News.*

"Cannot fail to be of wide interest, especially at the present time."—*Railway News.*

Pigments.

THE ARTIST'S MANUAL OF PIGMENTS. Showing their Composition, Conditions of Permanency, Non-Permanency, and Adulterations; Effects in Combination with Each Other and with Vehicles ; and the most Reliable Tests of Purity Together with the Science and Art Department's Examination Questions on Painting. By H. C. STANDAGE. Second Edition. Crown 8vo, 2s. 6d. cloth.

"This work is indeed *multum-in-parvo,* and we can, with good conscience, recommend it to all who come in contact with pigments, whether as makers, dealers or users."—*Chemical Review*

Gauging. Tables and Rules for Revenue Officers.

Brewers, etc.

A POCKET BOOK OF MENSURATION AND GAUGING : Containing Tables, Rules and Memoranda for Revenue Officers, Brewers, Spirit Merchants, &c. By J. B. MANT (Inland Revenue). 18mo, 4s. leather.

"This handy and useful book is adapted to the requirements of the Inland Revenue Department, and will be a favourite book of reference."—*Civilian.*

"Should be in the hands of every practical brewer."—*Brewers' Journal.*

INDUSTRIAL ARTS, TRADES, AND MANUFACTURES.

Cotton Spinning.

COTTON MANUFACTURE : A Manual of Practical Instruction in the Processes of Opening, Carding, Combing, Drawing, Doubling and Spinning of Cotton, the Methods of Dyeing, &c. For the Use of Operatives, Overlookers and Manufacturers. By JOHN LISTER, Technical Instructor, Pendleton. 8vo, 7s. 6d. cloth.   *Machinery,.*
"This invaluable volume is a distinct advance in the literature of cotton manufacture."—
"It is thoroughly reliable, fulfilling nearly all the requirements desired."—*Glasgow Herald.*

Flour Manufacture, Milling, etc.

FLOUR MANUFACTURE : A Treatise on Milling Science and Practice. By Professor FRIEDRICH KICK. Translated from the Second Enlarged and Revised Edition with Supplement. By H. H. P. POWLES, A.-M.Inst.C.E. Nearly 400 pp. Illustrated with 28 Folding Plates, and 167 Woodcuts. Royal 8vo, 25s. cloth.
"This valuable work is, and will remain, the standard authority on the science of milling. . . The miller who has read and digested this work will have laid the foundation, so to speak, of a successful career; he will have acquired a number of general principles which he can proceed to apply. In this handsome volume we at last have the accepted text-book of modern milling in good, sound English, which has little, if any, trace of the German idiom."—*The Miller.*

Agglutinants.

CEMENTS, PASTES, GLUES AND GUMS: A Practical Guide to the Manufacture and Application of the various Agglutinants required in the Building, Metal-Working, Wood-Working and Leather-Working Trades, and for Workshop, Laboratory or Office Use. With upwards of 900 Recipes and Formulæ. By H. C. STANDAGE. Crown 8vo, 2s. cloth.
"We have pleasure in speaking favourably of this volume. So far as we have had experience, which is not inconsiderable, this manual is trustworthy."—*Athenæum.*
"As a revelation of what are considered trade secrets, this book will arouse an amount of curiosity among the large number of industries it touches."—*Daily Chronicle.*

Soap-making.

THE ART OF SOAP-MAKING : A Practical Handbook of the *Manufacture of Hard and Soft Soaps, Toilet Soaps, etc.* By ALEXANDER WATT, Fifth Edition, Revised, including Modern Candle-Making. Crown 8vo, 7s.6d. cloth.   [*Just published.*
"The work will prove very useful, not merely to the technological student, but to the practical soap-boiler who wishes to understand the theory of his art."—*Chemical News.*
"A thoroughly practical treatise on an art which has almost no literature in our language. We congratulate the author on the success of his endeavour to fill a void in English technical literature."—*Nature.*

Paper Making.

PRACTICAL PAPER-MAKING : A Manual for Paper-makers and Owners and Managers of Paper-Mills. With Tables. Calculations, &c. By G. CLAPPERTON, Paper-maker. With Illustrations of Fibres from Micro-Photographs. Crown 8vo, 5s. cloth.   [*Just published.*
"The author caters for the requirements of responsible mill hands, apprentices, &c., whilst this manual will be found of great service to students of technology, as well as to veteran paper makers and mill owners. The illustrations form an excellent feature."—*Paper Trade Review.*
"We recommend everybody interested in the trade to get a copy of this thoroughly practical book."—*Paper Making.*

Paper Making.

THE ART OF PAPER MAKING : A Practical Handbook of the *Manufacture of Paper from Rags, Esparto, Straw, and other Fibrous Materials,* Including the Manufacture of Pulp from Wood Fibre, with a Description of the Machinery and Appliances used. To which are added Details of Processes for Recovering Soda from Waste Liquors. By ALEXANDER WATT, Author of " The Art of Soap-Making" With Illusts. Crown 8vo, 7s. 6d. cloth.
"It may be regarded as the standard work on the subject. The book is full of valuable information. The 'Art of Paper-making,' is in every respect a model of a text-book, either for a technical class or for the private student."—*Paper and Printing Trades Journal.*

Leather Manufacture.

THE ART OF LEATHER MANUFACTURE. Being a Practical Handbook, in which the Operations of Tanning, Currying, and Leather Dressing are fully Described, and the Principles of Tanning Explained, and many Recent Processes Introduced ; as also the Methods for the Estimation of Tannin, and a Description of the Arts of Glue Boiling, Gut Dressing, &c. By ALEXANDER WATT. Crown 8vo, 9s. cloth.
"A sound, comprehensive treatise on tanning and its accessories. It is an eminently valuable production, which redounds to the credit of both author and publishers."—*Chemical Review.*

Watch Adjusting.

THE WATCH ADJUSTER'S MANUAL : A Practical Guide for the Watch and Chronometer Adjuster in Making, Springing, Timing and Adjusting for Isochronism, Positions and Temperatures. By C. E. FRITTS. 370 pages, with Illustrations, 8vo, 16s. cloth. [Just published.

Horology.

A TREATISE ON MODERN HOROLOGY, in Theory and Practice. Translated from the French of CLAUDIUS SAUNIER, ex-Director of the School of Horology at Maçon, by JULIEN TRIPPLIN, F.R.A.S., Besançon Watch Manufacturer, and EDWARD RIGG, M.A., Assayer in the Royal Mint. With 78 Woodcuts and 22 Coloured Copper Plates. Second Edition. Super-royal 8vo, £2 2s. cloth ; £2 10s. half-calf.

"There is no horological work in the English language at all to be compared to this production of M. Saunier's for clearness and completeness. It is alike good as a guide for the student and as a reference for the experienced horologist and skilled workman."—*Horological Journal.*
"The latest, the most complete, and the most reliable of those literary productions to which continental watchmakers are indebted for the mechanical superiority over their English brethren —in tact, the Book of Books, is M. Saunier's 'Treatise.'"—*Watchmaker, Jeweller and Silversmith.*

Watchmaking.

THE WATCHMAKER'S HANDBOOK. Intended as a Workshop Companion for those engaged in Watchmaking and the Allied Mechanical Arts. Translated from the French of CLAUDIUS SAUNIER, and considerably enlarged by JULIEN TRIPPLIN, F.R.A.S., Vice-President of the Horological Institute, and EDWARD RIGG, M.A., Assayer in the Royal Mint. With numerous Woodcuts and 14 Copper Plates. Third Edition. Crown 8vo, 9s. cloth.

"Each part is truly a treatise in itself. The arrangement is good and the language is clear and concise. It is an admirable guide for the young watchmaker."—*Engineering.*
"It is impossible to speak too highly of its excellence. It fulfils every requirement in a handbook intended for the use of a workman. Should be found in every workshop."—*Watch and Clockmaker.*
"This book contains an immense number of practical details bearing on the daily occupation of a watchmaker."—*Watchmaker and Metalworker* (Chicago).

Watches and Timekeepers.

A HISTORY OF WATCHES AND OTHER TIMEKEEPERS. By JAMES F. KENDAL, M.B.H.Inst. 1s. 6d. boards ; or 2s. 6d. cloth gilt.

"Mr. Kendal's book, for its size, is the best which has yet appeared on this subject in the English language."—*Industries.*
"Open the book where you may, there is interesting matter in it concerning the ingenious devices of the ancient or modern horologer. The subject is treated in a liberal and entertaining spirit, as might be expected of a historian who is a master of the craft."—*Saturday Review.*

Electrolysis of Gold, Silver, Copper, etc.

ELECTRO-DEPOSITION : A Practical Treatise on the Electrolysis of Gold, Silver, Copper, Nickel, and other Metals and Alloys. With descriptions of Voltaic Batteries, Magneto and Dynamo-Electric Machines, Thermopiles, and of the Materials and Processes used in every Department of the Art, and several Chapters on Electro-Metallurgy. By ALEXANDER WATT, Author of "Electro-Metallurgy," &c. Third Edition, Revised. Crown 8vo, 9s. cloth.

"Eminently a book for the practical worker in electro-deposition. It contains practical descriptions of methods, processes and materials as actually pursued and used in the workshop."—*Engineer.*

Electro-Metallurgy.

ELECTRO-METALLURGY ; Practically Treated. By ALEXANDER WATT, Author of "Electro-Deposition," &c. Tenth Edition, including the most recent Processes. 12mo, 4s. cloth boards.

"From this book both amateur and artisan may learn everything necessary for the successful prosecution of electroplating."—*Iron.*

Working in Gold.

THE JEWELLER'S ASSISTANT IN THE ART OF WORKING IN GOLD : A Practical Treatise for Masters and Workmen. Compiled from the Experience of Thirty Years' Workshop Practice. By GEORGE E. GEE, Author of "The Goldsmith's Handbook," &c. Cr. 8vo, 7s. 6d. cloth.

"This manual of technical education is apparently destined to be a valuable auxiliary to a handicraft which is certainly capable of great improvement."—*The Times.*
"Very useful in the workshop, as the knowledge is practical, having been acquired by long experience, and all the recipes and directions are guaranteed to be successful."—*Jeweller and Metalworker.*

Electroplating.

ELECTROPLATING: A Practical Handbook on the Deposition of Copper, Silver, Nickel, Gold, Aluminium, Brass, Platinum, &c. &c. With Descriptions of the Chemicals, Materials, Batteries, and Dynamo Machines used in the Art. By J.W.Urquhart,C.E. Third Edition. Cr.8vo,5s.
"An excellent work, giving the newest information."—*Horological Journal.*

Electrotyping.

ELECTROTYPING . *The Reproduction and Multiplication of Printing Surfaces and Works of Art by the Electro-deposition of Metals.* By J. W. Urquhart, C.E. Crown 8vo, 5s. cloth.
"The book is thoroughly practical. The reader is, therefore, conducted through the leading laws of electricity, then through the metals used by electrotypers, the apparatus, and the depositing processes, up to the final preparation of the work,"—*Art Journal.*

Goldsmiths' Work.

THE GOLDSMITH'S HANDBOOK. By George E. Gee, Jeweller, &c. Third Edition, considerably Enlarged. 12mo, 3s. 6d. cl. bds.
"A good, sound educator, which will be accepted as an authority."—*Horological Journal.*

Silversmiths' Work.

THE SILVERSMITH'S HANDBOOK. By George E. Gee, Jeweller, &c. Second Edition, Revised. 12mo, 3s. 6d. cloth.
"The chief merit of the work is its practical character. . . The workers in the trade will speedily discover its merits when they sit down to study it."—*English Mechanic.*
⁎ *The above two works together, strongly half-bound, price 7s.*

Sheet Metal Working.

SHEET METAL WORKER'S INSTRUCTOR : Comprising a Selection of Geometrical Problems and Practical Rules for Describing the Various Patterns Required by Zinc, Sheet-Iron, Copper, and Tin-Plate Workers. By Reuben Henry Warn, Practical Tin-Plate Worker. New Edition, Revised and greatly Enlarged by Joseph G. Horner, A.M.I.M.E., Author of "Pattern Making," &c. Crown 8vo, 254 pages, with 430 Illustrations. 7s. 6d., cloth.   [*Just published.*

Bread and Biscuit Baking.

THE BREAD AND BISCUIT BAKER'S AND SUGAR-BOILER'S ASSISTANT. Including a large variety of Modern Recipes. By Robert Wells, Practical Baker. Crown 8vo, 2s. cloth.
"A large number of wrinkles for the ordinary cook, as well as the baker."—*Saturday Review.*

Confectionery for Hotels and Restaurants.

THE PASTRYCOOK AND CONFECTIONER'S GUIDE. For Hotels, Restaurants and the Trade in general, adapted also for Family Use. By Robert Wells. Crown 8vo, 2s. cloth.
"We cannot speak too highly of this really excellent work. In these days of keen competition our readers cannot do better than purchase this book."—*Bakers' Times.*

Ornamental Confectionery.

ORNAMENTAL CONFECTIONERY: A Guide for Bakers, Confectioners and Pastrycooks; including a variety of Modern Recipes, and Remarks on Decorative and Coloured Work. With 129 Original Designs. By Robert Wells. Crown 8vo, cloth gilt, 5s.
"A valuable work, practical, and should be in the hands of every baker and confectioner. The illustrative designs are alone worth treble the amount charged for the whole work."—*Bakers' Times.*

Flour Confectionery.

THE MODERN FLOUR CONFECTIONER. Wholesale and Retail. Containing a large Collection of Recipes for Cheap Cakes, Biscuits, &c. With Remarks on the Ingredients used in their Manufacture. By R. Wells. Crown 8vo, 2s. cloth.

Laundry Work.

LAUNDRY MANAGEMENT. A Handbook for Use in Private and Public Laundries, Including Descriptive Accounts of Modern Machinery and Appliances for Laundry Work. Small crown 8vo, 2s. cloth.
"This book should certainly occupy an honoured place on the shelves of all housekeepers who wish to keep themselves au courant of the newest appliances and methods."—*The Queen.*

HANDYBOOKS FOR HANDICRAFTS.

By PAUL N. HASLUCK,

EDITOR OF "WORK" (NEW SERIES); AUTHOR OF "LATHEWORK," "MILLING MACHINES," &c.

Crown 8vo, 144 pages, cloth, price 1s. each.

☞ *These* HANDYBOOKS *have been written to supply information for* WORKMEN, STUDENTS, *and* AMATEURS *in the several Handicrafts, on the actual* PRACTICE *of the* WORKSHOP, *and are intended to convey in plain language* TECHNICAL KNOWLEDGE *of the several* CRAFTS. *In describing the processes employed, and the manipulation of material, workshop terms are used; workshop practice is fully explained: and the text is freely illustrated with drawings of modern tools, appliances, and processes.*

THE METAL TURNER'S HANDYBOOK. A Practical Manual for Workers at the Foot-Lathe. With over 100 Illustrations. Price 1s.
" The book will be of service alike to the amateur and the artisan turner. It displays thorough knowledge of the subject."—*Scotsman.*

THE WOOD TURNER'S HANDYBOOK. A Practical Manual for Workers at the Lathe. With over 100 Illustrations. Price 1s.
" We recommend the book to young turners and amateurs. A multitude of workmen have hitherto sought in vain for a manual of this special industry."—*Mechanical World.*

THE WATCH JOBBER'S HANDYBOOK. A Practical Manual on Cleaning, Repairing, and Adjusting. With upwards of 100 Illustrations. Price 1s.
" We strongly advise all young persons connected with the watch trade to acquire and study this inexpensive work."—*Clerkenwell Chronicle.*

THE PATTERN MAKER'S HANDYBOOK. A Practical Manual on the Construction of Patterns for Founders. With upwards of 100 Illustrations. Price 1s.
" A most valuable, if not indispensable, manual for the pattern maker."—*Knowledge.*

THE MECHANIC'S WORKSHOP HANDYBOOK. A Practical Manual on Mechanical Manipulation. Embracing Information on various Handicraft Processes, with Useful Notes and Miscellaneous Memoranda. Comprising about 200 Subjects. Price 1s.
" A very clever and useful book, which should be found in every workshop; and it should certainly find a place in all technical schools."—*Saturday Review.*

THE MODEL ENGINEER'S HANDYBOOK. A Practical Manual on the Construction of Model Steam Engines. With upwards of 100 Illustrations. Price 1s.
" Mr. Hasluck has produced a very good little book."—*Builder.*

THE CLOCK JOBBER'S HANDYBOOK. A Practical Manual on Cleaning, Repairing, and Adjusting. With upwards of 100 Illustrations. Price 1s.
" It is of inestimable service to those commencing the trade."—*Coventry Standard.*

THE CABINET WORKER'S HANDYBOOK: A Practical Manual on the Tools, Materials, Appliances, and Processes employed in Cabinet Work. With upwards of 100 Illustrations. Price 1s.
" Mr. Hasluck's thoroughgoing little Handybook is amongst the most practical guides we have seen for beginners in cabinet-work."—*Saturday Review.*

THE WOODWORKER'S HANDYBOOK OF MANUAL INSTRUCTION. Embracing Information on the Tools, Materials, Appliances and Processes employed in Woodworking. With 104 Illustrations. Price 1s
[*Just published.*]

THE METALWORKER'S HANDYBOOK. With upwards of 100 Illustrations.
[*In preparation.*]

⁎⁎⁎ OPINIONS OF THE PRESS.

" Written by a man who knows, not only how work ought to be done, but how to do it, and how to convey his knowledge to others."—*Engineering.*
" Mr. Hasluck writes admirably, and gives complete instructions."—*Engineer.*
" Mr. Hasluck combines the experience of a practical teacher with the manipulative skill and scientific knowledge of processes of the trained mechanician, and the manuals are marvels of what can be produced at a popular price."—*Schoolmaster.*
" Helpful to workmen of all ages and degrees of experience."—*Daily Chronicle.*
" Practical, sensible, and remarkably cheap."—*Journal of Education.*
" Concise, clear and practical."—*Saturday Review.*

COMMERCE, COUNTING-HOUSE WORK, TABLES, etc.

Commercial French.

A NEW BOOK OF COMMERCIAL FRENCH: Grammar—Vocabulary — Correspondence — Commercial Documents — Geography — Arithmetic—Lexicon. By P. CARROUÉ, Professor in the City High School J.—B. Say (Paris). Crown 8vo, 4s. 6d. cloth.   [Just published.

Commercial Education.

LESSONS IN COMMERCE. By Professor R. GAMBARO, of the Royal High Commercial School at Genoa. Edited and Revised by JAMES GAULT, Professor of Commerce and Commercial Law in King's College, London. Second Edition, Revised. Crown 8vo, 3s. 6d. cloth.[Just published.

"The publishers of this work have rendered considerable service to the cause of commercial education by the opportune production of this volume. . . . The work is peculiarly acceptable to English readers and an admirable addition to existing class-books. In a phrase, we think the work attains its object in furnishing a brief account of those laws and customs of British trade with which the commercial man interested therein should be familiar."—Chamber of Commerce Journal.
"An invaluable guide in the hands of those who are preparing for a commercial career."
                                                                        Counting House.

Foreign Commercial Correspondence.

THE FOREIGN COMMERCIAL CORRESPONDENT: Being Aids to Commercial Correspondence in Five Languages—English, French, German, Italian, and Spanish. By CONRAD E. BAKER. Second Edition. Crown 8vo, 3s. 6d. cloth.

"Whoever wishes to correspond in all the languages mentioned by Mr. Baker cannot do better than study this work, the materials of which are excellent and conveniently arranged. They consist not of entire specimen letters but—what are far more useful—short passages, sentences, or phrases expressing the same general idea in various forms."—Athenæum.
"A careful examination has convinced us that it is unusually complete, well arranged, and reliable. The book is a thoroughly good one."—Schoolmaster.

Accounts for Manufacturers.

FACTORY ACCOUNTS: Their Principles and Practice. A Handbook for Accountants and Manufacturers, with Appendices on the Nomenclature of Machine Details; the Income Tax Acts; the Rating of Factories; Fire and Boiler Insurance; the Factory and Workshop Acts, &c., including also a Glossary of Terms and a large number of Specimen Rulings. By EMILE GARCKE and J. M. FELLS. Fourth Edition, Revised and Enlarged. Demy 8vo, 250 pages, 6s. strongly bound.

"A very interesting description of the requirements of Factory Accounts. . . . the principle of assimilating the Factory Accounts to the general commercial books is one which we thoroughly agree with."—Accountants' Journal.
"Characterised by extreme thoroughness. There are few owners of factories who would not derive great benefit from the perusal of this most admirable work."—Local Government Chronicle.

Modern Metrical Units and Systems.

MODERN METROLOGY: A Manual of the Metrical Units and Systems of the Present Century. With an Appendix containing a proposed English System. By LOWIS D'A. JACKSON, A.M.Inst.C.E., Author of "Aid to Survey Practice," &c. Large crown 8vo 12s. 6d. cloth.

"We recommend the work to all interested in the practical reform of our weights and measures."—Nature.

The Metric System and the British Standards.

A SERIES OF METRIC TABLES, in which the British Standard Measures and Weights are compared with those of the Metric System at present in Use on the Continent. By C. H. DOWLING, C.E. 8vo, 10s. 6d. strongly bound.

"Mr. Dowling's Tables are well put together as a ready-reckoner for the conversion of one system into the other."—Athenæum.

Iron Shipbuilders' and Merchants' Weight Tables.

IRON-PLATE WEIGHT TABLES: For Iron Shipbuilders, Engineers, and Iron Merchants. Containing the Calculated Weights of upwards of 150,000 different sizes of Iron Plates, from 1 foot by 6 in. by ¼ in. to 10 feet by 5 feet by 1 in. Worked out on the basis of 40 lbs. to the square foot of Iron of 1 inch in thickness. Carefully compiled and thoroughly Revised by H. BURLINSON and W. H. SIMPSON. Oblong 4to, 25s. half-bound.

"This work will be found of great utility. The authors have had much practical experience of what is wanting in making estimates; and the use of the book will save much time in making elaborate calculations."—English Mechanic.

Chadwick's Calculator for Numbers and Weights Combined.

THE NUMBER, WEIGHT, AND FRACTIONAL CALCU-LATOR. Containing upwards of 250,000 Separate Calculations, showing at a glance the value at 422 different rates, ranging from $\frac{1}{16}$th of a Penny to 20s. each, or per cwt., and £20 per ton, of any number of articles consecutively, from 1 to 470.—Any number of cwts., qrs., and lbs., from 1 cwt. to 470 cwts.—Any number of tons, cwts., qrs., and lbs., from 1 to 1,000 tons. By WILLIAM CHADWICK, Public Accountant. Third Edition, Revised and Improved. 8vo, 18s., strongly bound for Office wear and tear.

☞ *Is adapted for the use of Accountants and Auditors, Railway Companies, Canal Companies, Shippers, Shipping Agents, General Carriers, etc. Ironfounders, Brassfounders, Metal Merchants, Iron Manufacturers, Ironmongers, Engineers, Machinists, Boiler Makers, Millwrights, Roofing, Bridge and Girder Makers, Colliery Proprietors, etc. Timber Merchants, Builders, Contractors, Architects, Surveyors, Auctioneers, Valuers, Brokers, Mill Owners and Manufacturers, Mill Furnishers, Merchants, and General Wholesale Tradesmen. Also for the Apportionment of Mileage Charges for Railway Traffic.*

"It is as easy of reference for any answer or any number of answers as a dictionary, and the references are even more quickly made. For making up accounts or estimates the book must prove invaluable to all who have any considerable quantity of calculations involving price and measure in any combination to do."—*Engineer.*

Harben's Comprehensive Weight Calculator.

THE WEIGHT CALCULATOR. Being a Series of Tables upon a New and Comprehensive Plan, exhibiting at One Reference the exact Value of any Weight from 1 lb. to 15 tons, at 300 Progressive Rates, from 1d. to 168s. per cwt., and containing 186,000 Direct Answers, which, with their Combinations, consisting of a single addition (mostly to be performed at sight), will afford an aggregate of 10,266,000 Answers; the whole being calculated and designed to ensure correctness and promote despatch. By HENRY HARBEN, Accountant. Fifth Edition, carefully Corrected Royal 8vo, £1 5s. strongly half-bound. [*Just published.*]

"A practical and useful work of reference for men of business generally; it is the best of the kind we have seen."—*Ironmonger.*

"Of priceless value to business men. It is a necessary book in all mercantile offices."—*Sheffield Independent.*

Harben's Comprehensive Discount Guide.

THE DISCOUNT GUIDE. Comprising several Series of Tables for the use of Merchants, Manufacturers, Ironmongers, and others, by which may be ascertained the exact Profit arising from any mode of using Discounts, either in the Purchase or Sale of Goods, and the method of either Altering a Rate of Discount or Advancing a Price, so as to produce, by one operation, a sum that will realise any required profit after allowing one or more Discounts: to which are added Tables of Profit or Advance from 1¼ to 90 per cent., Tables of Discount from 1¼ to 98¼ per cent., and Tables of Commission, &c., from ¼ to 10 per cent. By HENRY HARBEN, Accountant. New Edition, Revised and Corrected. Demy 8vo, 544 pp., £1 5s. half-bound.

"A book such as this can only be appreciated by business men, to whom the saving of time means saving of money. We have the high authority of Professor J. R. Young that the tables throughout the work are constructed upon strictly accurate principles. The work is a model of typographical clearness, and must prove of great value to merchants, manufacturers, and general traders."—*British Trade Journal.*

New Wages Calculator.

TABLES OF WAGES at 54, 52, 50 and 48 Hours per Week. Showing the Amounts of Wages from One-quarter-of-an hour to Sixty-four hours in each case at Rates of Wages advancing by One Shilling from 4s. to 55s. per week. By THOS. GARBUTT, Accountant. Square crown 8vo, 6s. half-bound. [*Just published.*]

Iron and Metal Trades' Calculator.

THE IRON AND METAL TRADES' COMPANION. For expeditiously ascertaining the Value of any Goods bought or sold by Weight, from 1s. per cwt. to 112s. per cwt., and from one farthing per pound to one shilling per pound. By THOMAS DOWNIE. 396 pp., 9s. leather.

"A most useful set of tables; nothing like them before existed."—*Building News.*

"Although specially adapted to the iron and metal trades, the tables will be found useful in every other business in which merchandise is bought and sold by weight."—*Railway News.*

"DIRECT CALCULATORS,"

By M. B. COTSWORTH, of Holgate, York.

QUICKEST AND MOST ACCURATE MEANS OF CALCULATION KNOWN.
ENSURE ACCURACY and SPEED WITH EASE, SAVE TIME and MONEY.
Accounts may be charged out or checked by these means in about one
third he time required by ordinary methods of calculation. These
unrivalled "Calculators" have very clear and original contrivances
for instantly finding the exact answer, by its fixed position, without
even sighting the top or side of the page. They are varied in arrangement to suit the special need of each particular trade.

All the leading firms now use Calculators, even where they employ experts.

N.B.—Indicator letters in brackets should be quoted.

"*RAILWAY & TRADERS' CALCULATOR* " (**R. & T.**) 10s. 6d.
Including Scale of Charges for Small Parcels by Merchandise Trains.
"Direct Calculator"—the only Calculator published giving exact charge for
Cwts., Qrs. and Lbs., together. "Calculating Tables" for every 1d. rate to
100s. per ton. "Wages Calculator." "Percentage Rates." "Grain, Flour,
Ale, &c., Weight Calculators."

"*DIRECT CALCULATOR* (**I R**) " including all the above except
"Calculating Tables." 7s.

"*DIRECT CALCULATOR* (**A**) " by ½d., 2s. each opening, exact
pence to 40s. per ton. 5s.

"*DIRECT CALCULATOR* (**B**) " by 1d., 4s. each opening, exact
pence to 40s. per ton. 4s. 6d.

"*DIRECT CALCULATOR* (**C**) " by 1d. (with Cwts. and Qrs. to
nearest farthing), to 40s. per ton. 4s. 6d.

"*DIRECT CALCULATOR* (**Ds**) " by 1d. gradations. (Single Tons
to 50 Tons, then by fifties to 1,000 Tons, with Cwts. values below in exact
pence payable, fractions of ½d. and upwards being counted as 1d. 6s. 6d.

"*DIRECT CALCULATOR* (**D**) " has from 1,000 to 10,000 Tons in
addition to the (**Ds**) Calculator. 7s. 6d.

"*DIRECT CALCULATOR* (**Es**) " by 1d. gradations. (As (**D**) to
1,000 Tons, with Cwts. and Qrs. values shown separately to the nearest
farthing). 5s. 6d.

"*DIRECT CALCULATOR* (**E**) " has from 1,000 to 10,000 Tons in
addition to the (**Es**) Calculator. 6s. 6d.

"*DIRECT CALCULATOR* (**F**) " by 1d., 2s. each opening, exact
pence to 40s. per ton. 4s. 6d.

"*DIRECT CALCULATOR* (**G**) " by 1d., 1s. each opening; 6 in. by
9 in. Nearest ¼d. Indexed (**G I**) 3s. 6d. 2s. 6d.

"*DIRECT CALCULATOR* (**H**) " by 1d., 1s. each opening; 6 in. by
9 in. To exact pence. Indexed (**H I**) 3s. 6d. 2s. 6d.

"*DIRECT CALCULATOR* (**K**) " Showing Values of Tons, Cwts.
and Qrs. in even pence (fractions of 1d. as 1d.), for the Retail Coal Trade.
4s. 6d.

"*RAILWAY AND TIMBER TRADES MEASURER AND CAL-
CULATOR* (**T**) " (as prepared for the Railway Companies). The only book
published giving true content of unequal sided and round timber by eighths
of an inch, quarter girth, Weights from Cubic Feet—Standards, Superficial
Feet, and Stone to Weights—Running Feet from lengths of Deals—Standard
Multipliers—Timber Measures—Customs Regulations, &c. 3s. 6d.

AGRICULTURE, FARMING, GARDENING, etc.

Dr. Fream's New Edition of "The Standard Treatise on Agriculture."

THE COMPLETE GRAZIER, and FARMER'S and CATTLE-BREEDER'S ASSISTANT: A Compendium of Husbandry. Originally Written by WILLIAM YOUATT. Thirteenth Edition, entirely Re-written, considerably Enlarged, and brought up to the Present Requirements of Agricultural Practice, by WILLIAM FREAM, LL.D., Steven Lecturer in the University of Edinburgh, Author of "The Elements of Agriculture," &c. Royal 8vo, 1,100 pp., with over 450 Illustrations. £1 11s. 6d. strongly and handsomely bound.

EXTRACT FROM PUBLISHERS' ADVERTISEMENT.

"A treatise that made its original appearance in the first decade of the century, and that enters upon its Thirteenth Edition before the century has run its course, has undoubtedly established its position as a work of permanent value. . . The phenomenal progress of the last dozen years in the Practice and Science of Farming has rendered it necessary, however, that the volume should be re-written, . . . and for this undertaking the publishers were fortunate enough to secure the services of Dr. FREAM, whose high attainments in all matters pertaining to agriculture have been so emphatically recognised by the highest professional and official authorities. In carrying out his editorial duties, Dr. FREAM has been favoured with valuable contributions by Prof. J. WORTLEY AXE, Mr. E. BROWN, Dr. BERNARD DYER, Mr. W. J. MALDEN, Mr. R. H. REW, Prof. SHELDON, Mr. J. SINCLAIR, Mr. SANDERS SPENCER, and others.

"As regards the illustrations of the work, no pains have been spared to make them as representative and characteristic as possible, so as to be practically useful to the Farmer and Grazier."

SUMMARY OF CONTENTS.

BOOK I. ON THE VARIETIES, BREEDING, REARING, FATTENING, AND MANAGEMENT OF CATTLE.

BOOK II. ON THE ECONOMY AND MANAGEMENT OF THE DAIRY.

BOOK III. ON THE BREEDING, REARING, AND MANAGEMENT OF HORSES.

BOOK IV. ON THE BREEDING, REARING, AND FATTENING OF SHEEP.

BOOK V. ON THE BREEDING, REARING, AND FATTENING OF SWINE.

BOOK VI. ON THE DISEASES OF LIVE STOCK.

BOOK VII. ON THE BREEDING, REARING, AND MANAGEMENT OF POULTRY.

BOOK VIII. ON FARM OFFICES AND IMPLEMENTS OF HUSBANDRY.

BOOK IX. ON THE CULTURE AND MANAGEMENT OF GRASS LANDS.

BOOK X. ON THE CULTIVATION AND APPLICATION OF GRASSES, PULSE, AND ROOTS.

BOOK XI. ON MANURES AND THEIR APPLICATION TO GRASS LAND & CROPS

BOOK XII. MONTHLY CALENDARS OF FARMWORK.

**** OPINIONS OF THE PRESS ON THE NEW EDITION.

" Dr. Fream is to be congratulated on the successful attempt he has made to give us a work which will at once become the standard classic of the farm practice of the country. We believe that it will be found that it has no compeer among the many works at present in existence. . . . The illustrations are admirable, while the frontispiece, which represents the well-known bull, New Year's Gift, bred by the Queen, is a work of art."—*The Times.*

" The book must be recognised as occupying the proud position of the most exhaustive work of reference in the English language on the subject with which it deals."—*Athenæum.*

" The most comprehensive guide to modern farm practice that exists in the English language to-day. . . . The book is one that ought to be on every farm and in the library of every landowner."—*Mark Lane Express.*

" In point of exhaustiveness and accuracy the work will certainly hold a pre-eminent and unique position among books dealing with scientific agricultural practice. It is, in fact, an agricultural library of itself."— *North British Agriculturist.*

" A compendium of authoritative and well-ordered knowledge on every conceivable branch of the work of the live stock farmer ; probably without an equal in this or any other country."

Yorkshire Post.

British Farm Live Stock.

FARM LIVE STOCK OF GREAT BRITAIN. By ROBERT WALLACE, F.L.S., F.R.S.E., &c., Professor of Agriculture and Rural Economy in the University of Edinburgh. Third Edition, thoroughly Revised and considerably Enlarged. With over 120 Phototypes of Prize Stock. Demy 8vo. 384 pp., with 79 Plates and Maps, 12s. 6d. cloth.

" A really complete work on the history, breeds, and management of the farm stock of Great Britain, and one which is likely to find its way to the shelves of every country gentleman's library."—*The Times.*

" The latest edition of ' Farm Live Stock of Great Britain ' is a production to be proud of, and its issue not the least of the services which its author has rendered to agricultural science."

Scottish Farmer.

" The book is very attractive . . . and we can scarcely imagine the existence of a farmer who would not like to have a copy of this beautiful work."—*Mark Lane Express.*

" A work which will long be regarded as a standard authority whenever a concise history and description of the breeds of live stock in the British Isles is required."—*Bell's Weekly Messenger.*

Dairy Farming.

BRITISH DAIRYING. A Handy Volume on the Work of the Dairy-Farm. For the Use of Technical Instruction Classes, Students in Agricultural Colleges, and the Working Dairy-Farmer. By Prof. J. P. SHELDON, ate Special Commissioner of the Canadian Government, Author of " Dairy Farming," &c. With numerous Illustrations. Crown 8vo, 2s. 6d. cloth.

" We confidently recommend it as a text-book on dairy farming."—*Agricultural Gazette.*
" Probably the best half-crown manual on dairy work that has yet been produced."—*North British Agriculturist.*
" It is the soundest little work we have yet seen on the subject."—*The Times.*

Dairy Manual.

MILK, CHEESE AND BUTTER: A Practical Handbook on their Properties and the Processes of their Production, including a Chapter on Cream and the Methods of its Separation from Milk. By JOHN OLIVER, late Principal of the Western Dairy Institute, Berkeley. With Coloured Plates and 200 Illusts. Crown 8vo, 7s. 6d. cloth.   [*Just published.*

" An exhaustive and masterly production. It may be cordially recommended to all students and practitioners of dairy science."—*N.B. Agriculturist.*
" We strongly recommend this very comprehensive and carefully-written book to dairy-farmers and students of dairying. It is a distinct acquisition to the library of the agriculturist."—*Agricultural Gazette.*

Agricultural Facts and Figures.

NOTE-BOOK OF AGRICULTURAL FACTS AND FIGURES FOR FARMERS AND FARM STUDENTS. By PRIMROSE McCONNELL, B.Sc. Fifth Edition. Royal 32mo. roan, gilt edges, with band, 4s.

" Literally teems with information, and we can cordially recommend it to all connected with agriculture."—*North British Agriculturist.*

Small Farming.

SYSTEMATIC SMALL FARMING; or, The Lessons of my Farm. Being an Introduction to Modern Farm Practice for Small Farmers. By R. SCOTT BURN. With numerous Illustrations, crown 8vo, 6s. cloth.

" This is the completest book of its class we have seen, and one which every amateur farmer will read with pleasure and accept as a guide."—*Field.*

Modern Farming.

OUTLINES OF MODERN FARMING. By R. SCOTT BURN. Soils, Manures, and Crops—Farming and Farming Economy—Cattle, Sheep, and Horses — Management of Dairy, Pigs, and Poultry — Utilisation of Town-Sewage, Irrigation, &c. Sixth Edition. In One Vol., 1,250 pp., half-bound, profusely Illustrated, 12s.

" The aim of the author has been to make his work at once comprehensive and trustworthy and he has succeeded to a degree which entitles him to much credit."—*Morning Advertiser.*

Agricultural Engineering.

FARM ENGINEERING, THE COMPLETE TEXT-BOOK OF. Comprising Draining and Embanking; Irrigation and Water Supply; Farm Roads, Fences, and Gates; Farm Buildings; Barn Implements and Machines; Field Implements and Machines; Agricultural Surveying, &c. By Prof JOHN SCOTT. 1,150 pages, half-bound, with over 600 Illustrations, 12s.

" Written with great care, as well as with knowledge and ability. The author has done his work well: we have found him a very trustworthy guide wherever we have tested his statements. The volume will be of great value to agricultural students."—*Mark Lane Express.*

Agricultural Text-Book.

THE FIELDS OF GREAT BRITAIN: A Text-Book of Agriculture, adapted to the Syllabus of the Science and Art Department. For Elementary and Advanced Students. By HUGH CLEMENTS (Board of Trade). Second Edition, Revised, with Additions. 18mo, 2s. 6d. cloth.

" A most comprehensive volume, giving a mass of information."—*Agricultural Economist.*
" It is a long time since we have seen a book which has pleased us more, or which contains such a vast and useful fund of knowledge."—*Educational Times.*

Tables for Farmers, etc.

TABLES, MEMORANDA, AND CALCULATED RESULTS for Farmers. Graziers, Agricultural Students, Surveyors, Land Agents, Auctioneers, etc. With a New System of Farm Book-keeping. By SIDNEY FRANCIS. Third Edition, Revised. 272 pp., waistcoat-pocket size, 1s. 6d. leather.

" Weighing less than 1 oz., and occupying no more space than a match box, it contains a mass of facts and calculations which has never before, in such handy form, been obtainable. Every operation on the farm is dealt with. The work may be taken as thoroughly accurate, the whole of the tables having been revised by Dr. Fream. We cordially recommend it."—*Bell's Weekly Messenger.*

Artificial Manures and Foods.

FERTILISERS AND FEEDING STUFFS : Their Proper-
ties and Uses. A Handbook for the Practical Farmer. By BERNARD DYER,
D.Sc. (Lond.) With the Text of the Fertilisers and Feeding Stuffs Act of
1893, the Regulations and Forms of the Board of Agriculture and Notes on
the Act by A. J. DAVID, B.A., LL.M., of the Inner Temple, Barrister-at-Law.
Crown 8vo, 120 pages, 1s. cloth.   [*Just published.*
"An excellent shillingsworth. Dr. Dyer has done farmers good service in placing at their dis-
posal so much useful information in so intelligible a form."—*The Times.*

The Management of Bees.

BEES FOR PLEASURE AND PROFIT : A Guide to the
Manipulation of Bees, the Production of Honey, and the General Management of the Apiary. By G. GORDON SAMSON. Crown 8vo, 1s. cloth.
" The intending bee-keeper will find exactly the kind of Information required to enable him
to make a successful start with his hives. The author is a thoroughly competent teacher, and his
book may be commended."—*Morning Post.*

Farm and Estate Book-keeping.

BOOK-KEEPING FOR FARMERS & ESTATE OWNERS.
A Practical Treatise, presenting, in Three Plans, a System adapted for all
Classes of Farms. By JOHNSON M. WOODMAN, Chartered Accountant. Second
Edition, Revised. Crown 8vo, 3s. 6d. cloth boards ; or 2s. 6d. cloth limp.
"The volume is a capital study of a most important subject."—*Agricultural Gazette.*
The young farmer, land agent, and surveyor will find Mr. Woodman's treatise more than
repay its cost and study.'—*Building News.*

Farm Account Book.

WOODMAN'S YEARLY FARM ACCOUNT BOOK. Giving
a Weekly Labour Account and Diary, and showing the Income and Expen-
diture under each Department of Crops, Live Stock, Dairy, &c. &c. With
Valuation, Profit and Loss Account, and Balance Sheet at the end of the
Year. By JOHNSON M. WOODMAN, Chartered Accountant, Author of " Book-
keeping for Farmers." Folio, 7s. 6d. half bound.   [*culture.*
"Contains every requisite form for keeping farm accounts readily and accurately."—*Agri-*

Early Fruits, Flowers, and Vegetables.

THE FORCING GARDEN : or, How to Grow Early Fruits,
Flowers, and Vegetables. With Plans and Estimates for Building Glass-
houses, Pits, and Frames. By SAMUEL WOOD. Crown 8vo, 3s. 6d. cloth.
" A good book, and fairly fills a place that was in some degree vacant. The book is written with
great care, and contains a great deal of valuable teaching."—*Gardeners' Magazine.*

Good Gardening.

A PLAIN GUIDE TO GOOD GARDENING ; or, How to Grow
Vegetables, Fruits, and Flowers. By S. WOOD. Fourth Edition, with con-
siderable Additions, &c., and numerous Illustrations. Crown 8vo. 3s. 6d. cl.
" A very good book, and one to be highly recommended as a practical guide. The practical
directions are excellent."—*Athenæum.*
"May be recommended to young gardeners, cottagers, and specially to amateurs, for the
plain, simple, and trustworthy information it gives on common matters too often neglected."—
Gardeners' Chronicle.

Gainful Gardening.

MULTUM-IN-PARVO GARDENING ; or, How to make One
Acre of Land produce £620 a-year by the Cultivation of Fruits and Vegetables ;
also, How to Grow Flowers in Three Glass Houses, so as to realise £176 per
annum clear Profit. By SAMUEL WOOD. Author of " Good Gardening," &c.
Fifth and Cheaper Edition, Revised, with Additions. Crown 8vo, 1s. sewed.
" We are bound to recommend it as not only suited to the case of the amateur and gentleman's
gardener, but to the market grower."—*Gardeners' Magazine.*

Gardening for Ladies.

THE LADIES' MULTUM-IN-PARVO FLOWER GARDEN,
and Amateurs' Complete Guide. With Illusts. By S. WOOD. Cr. 8vo, 3s. 6d. cl.

Receipts for Gardeners.

GARDEN RECEIPTS. Edited by CHARLES W. QUIN. 12mo,
1s. 6d. cloth limp.

Market Gardening.

MARKET AND KITCHEN GARDENING. By Contributors
to "The Garden." Compiled by C. W. SHAW, late Editor of "Gardening
Illustrated." 12mo 3s. 6d. cloth boards.

AUCTIONEERING, VALUING, LAND SURVEYING ESTATE AGENCY, etc.

Auctioneer's Assistant.

THE APPRAISER, AUCTIONEER, BROKER, HOUSE AND ESTATE AGENT AND VALUER'S POCKET ASSISTANT, for the Valuation for Purchase, Sale, or Renewal of Leases, Annuities and Reversions, **and** of property generally; with Prices for Inventories, &c. By JOHN WHEELER, Valuer, &c. Sixth Edition, Re-written and greatly extended by C. NORRIS, Surveyor, Valuer, &c. Royal 32mo, 5s. cloth.

"A neat and concise book of reference, containing an admirable and clearly-arranged list of prices for inventories, and a very practical guide to determine the value of furniture, &c."—*Standard.*

"Contains a large quantity of varied and useful information as to the valuation for purchase, sale, or renewal of leases, annuities and reversions, and of property generally, with prices for inventories, and a guide to determine the value of interior fittings and other effects."—*Builder.*

Auctioneering.

AUCTIONEERS: THEIR DUTIES AND LIABILITIES. A Manual of Instruction and Counsel for the Young Auctioneer. By ROBERT SQUIBBS, Auctioneer. Second Edition, Revised and partly Re-written. Demy 8vo, 12s. 6d. cloth.

*** OPINIONS OF THE PRESS.

"The standard text-book on the topics of which it treats."—*Athenæum.*

"The work is one of general excellent character, and gives much information in a compendious and satisfactory form."—*Builder.*

"May be recommended as giving a great deal of information on the law relating to auctioneers, in a very readable form."—*Law Journal.*

"Auctioneers may be congratulated on having so pleasing a writer to minister to their special needs."—*Solicitors' Journal.*

"Every auctioneer ought to possess a copy of this excellent work."—*Ironmonger.*

"Of great value to the profession. . . . We readily welcome this book from the fact that it treats the subject in a manner somewhat new to the profession."—*Estates Gazette.*

Inwood's Estate Tables.

TABLES FOR THE PURCHASING OF ESTATES, Freehold, Copyhold, or Leasehold; Annuities, Advowsons, etc., and for the Renewing of Leases held under Cathedral Churches, Colleges, or other Corporate bodies for Terms of Years certain, and for Lives: also for Valuing Reversionary Estates, Deferred Annuities, Next Presentations, &c.; together with SMART'S Five Tables of Compound Interest, and an Extension of the same to Lower and Intermediate Rates. By W. INWOOD. 24th Edition, with considerable Additions, and new and valuable Tables of Logarithms for the more Difficult Computations of the Interest of Money, Discount, Annuities, &c., by M. FEDOR THOMAN, of the Société Crédit Mobilier of Paris. Crown 8vo, 8s. cloth.

"Those interested in the purchase and sale of estates, and in the adjustment of compensation cases, as well as in transactions in annuities, life insurances, &c., will find the present edition of eminent service."—*Engineering.*

"'Inwood's Tables' still maintain a most enviable reputation. The new issue has been enriched by arge additional contributions by M. Fedor Thoman, whose carefully arranged Tables cannot fail to be of the utmost utility."—*Mining Journal.*

Agricultural Valuer's Assistant.

THE AGRICULTURAL VALUER'S ASSISTANT. A Practical Handbook on the Valuation of Landed Estates; including Rules and Data for Measuring and Estimating the Contents, Weights, and Values of Agricultural Produce and Timber, and the Values of Feeding Stuffs, Manures, and Labour; with Forms of Tenant-Right-Valuations, Lists of Local Agricultural Customs, Scales of Compensation under the Agricultural Holdings Act, &c. &c. By TOM BRIGHT, Agricultural Surveyor. Second Edition, much Enlarged. Crown 8vo, 5s. cloth.

"Full of tables and examples in connection with the valuation of tenant-right, estates, labour, contents, and weights of timber, and farm produce of all kinds."—*Agricultural Gazette.*

"An eminently practical handbook, full of practical tables and data of undoubted interest and value to surveyors and auctioneers in preparing valuations of all kinds."—*Farmer.*

Plantations and Underwoods.

POLE PLANTATIONS AND UNDERWOODS: A Practical Handbook on Estimating the Cost of Forming, Renovating, Improving, and Grubbing Plantations and Underwoods, their Valuation for Purposes of Transfer, Rental, Sale, or Assessment. By TOM BRIGHT, Author of "The Agricultural Valuer's Assistant," &c. Crown 8vo, 3s. 6d. cloth.

"To valuers, foresters and agents it will be a welcome aid."—*North British Agriculturist.*

"Well calculated to assist the valuer in the discharge of his duties, and of undoubted interest and use both to surveyors and auctioneers in preparing valuations of all kinds."—*Kent Herald.*

www.ingramcontent.com/pod-product-compliance
Lightning Source LLC
Chambersburg PA
CBHW021806190326
41518CB00007B/470